Liquid Metals

Edited by Samson Jerold Samuel Chelladurai, S. Gnanasekaran and Suresh Mayilswamy

Published in London, United Kingdom

IntechOpen

Supporting open minds since 2005

Liquid Metals
http://dx.doi.org/10.5772/intechopen.92522
Edited by Samson Jerold Samuel Chelladurai, S. Gnanasekaran and Suresh Mayilswamy

Contributors
Bazani Shaik, Gosala Harinath Gowd, Bandaru Durga Prasad, Hadef Zakaria, Kamli Kenza, Xianfeng Xiao, Cong Lu, Yansu Fu, Xiaojun Ye, Lijun Song, Samson Jerold Samuel Chelladurai, S. Gnanasekaran, G. Padmanaban, S. Sivananthan, Ramesh S. Bhat

First published in London, United Kingdom, 2021 by IntechOpen
IntechOpen is the global imprint of INTECHOPEN LIMITED, registered in England and Wales, registration number: 11086078, 5 Princes Gate Court, London, SW7 2QJ, United Kingdom
Printed in Croatia

British Library Cataloguing-in-Publication Data
A catalogue record for this book is available from the British Library

Additional hard and PDF copies can be obtained from orders@intechopen.com

Liquid Metals
Edited by Samson Jerold Samuel Chelladurai, S. Gnanasekaran and Suresh Mayilswamy
p. cm.
Print ISBN 978-1-83969-184-3
Online ISBN 978-1-83969-185-0
eBook (PDF) ISBN 978-1-83969-186-7

Meet the editors

Dr. Samson Jerold Samuel Chelladurai, Ph.D., is an associate professor in the Department of Mechanical Engineering, Sri Krishna College of Engineering and Technology (SKCET), Coimbatore, Tamil Nadu, India. He received a Gold Medal and secured University First Rank for his outstanding performance in post-graduation. He obtained a Ph.D. in Composite Materials from Anna University, Chennai, India, in 2018. He is specialized in the manufacturing of metal matrix composites, characterization of composites, tribology, and optimization techniques. He is currently guiding six research scholars in the fields of manufacturing engineering. He has published many research articles in SCI- and Scopus-indexed journals. He served as lead editor for special issues in various international journals. He has also published five books, five patents, and received grants worth $89,000 USD. He has received several awards from SKCET and other organizations and societies.

Dr. S. Gnanasekaran, Ph.D., is an assistant professor in the Department of Mechanical Engineering, Sri Sakthi Institute of Engineering and Technology, Tamil Nadu, India. He obtained a Ph.D. in Welding and Surface Engineering. His research work was funded by the University Grants Commission – Department of Atomic Energy (UGC-DAE). He has published many research articles in SCI- and Scopus-indexed journals and published five patents. He has also published five book chapters. Dr. Gnanasekaran received several awards from international conferences for his technical work. He also received the best faculty award for his immeasurable contribution to students' welfare. He has been an editorial board member, advisory board member, and reviewer for several journals.

Dr. Suresh Mayilswamy, Ph.D., is an associate professor in the Department of Robotics and Automation Engineering, PSG College of Technology, India. His Ph.D. research focused on theoretical and experimental investigations of orientations of brake pads in vibratory part feeders. His current research focuses on advanced manufacturing technology, process automation, and industrial robotics. He has published around seventy-three technical papers in refereed and impact-factored international journals. He has authored three books in the field of industrial automation and assembly automation. He received the outstanding faculty award and best faculty of the year from SKCET and PSG Institute of Advanced Studies for his academic and research works in 2013 and 2018, respectively.

Contents

Preface

This book discusses liquid metals used in various manufacturing processes in the aerospace and automobile industries. It provides important original and theoretical experimental results on the use of non-routine technologies. It also presents novel applications of more familiar experimental techniques and analyses of composites. Topics covered include the importance of liquid metals, friction stir welding to improve aluminium alloys, adhesion phenomenon of liquid metals, secondary aluminium used for producing products, high-temperature wear characteristics of plasma transferred arc hardfaced Ni–Cr–Si–B-C alloy deposits, characteristics of an aluminum-silicon alloy in liquid state during the centrifugal casting process, and melt pool convection during laser material processing.

The editors are grateful to their management, institutions, family members and chapter contributors. Additionally, we like to thank IntechOpen for their tireless efforts throughout the preparation and publication of this book.

Dr. Samson Jerold Samuel Chelladurai
Associate Professor,
Mechanical Engineering,
Sri Krishna College of Engineering and Technology,
Coimbatore, Tamil Nadu, India

Dr. S. Gnanasekaran
Sri Shakthi Institute of Engineering and Technology,
India

Dr. Suresh Mayilswamy
PSG College of Technology,
India

Section 1

Welding

Chapter 1

Introductory Chapter: Emerging Trends in Liquid Metals

Samson Jerold Samuel Chelladurai

1. Introduction

Liquid metals play a vital role in manufacturing of products. The state of liquid metal influences the microstructure and mechanical properties of end properties. Hardness, tensile strength, impact strength, fatigue strength, wear resistance, corrosion resistance are influenced by the liquid metal during manufacturing processes. Industries focused to produce the components with superior properties by utilizing the liquid metal in an effective way.

Liquid metal used in casting process used to produce the components according to the shape and size of the die used. In this process, the metal is heated in a furnace and reinforcements are added to produce the homogeneous mixture. The liquid molten mixture is poured into the die to produce the castings. Stir casting and squeeze casting process normally employed to produce particle reinforced composites and fiber reinforced composites can be produced by squeeze casting process. The products produced by these processes provide excellent hardness, tensile strength and wear resistance.

In addition, friction stir welding process is used to join similar and dissimilar metals and alloys. In this process, the rotational tool is used and frictional force is used to produce the heat during joining process. The rotational speed, frictional force and tool geometry which influence the properties of aluminum alloy. Industries using friction stir welding process in automotive, aerospace and marine industries because of its better welding efficiency.

At present, friction stir processing is used to modify the properties of materials at surface level. The hardness of materials can be improved by reinforcing nano particles in base material. The reinforcements are placed in holes and the rotational tool is used to produce the surface composites. The products produced by friction stir processing exhibits better hardness, tensile strength, corrosion and resistance to wear.

2. Conclusion

The main aim of this collection of book chapters primarily focused on Liquid metals of various manufacturing process used in aerospace and automobile industries. This volume offers original and experimental results which use new technologies which make the readers to read it. Review and research book chapters present novel research work in the field of composites, welding techniques and aluminum alloy.

The book highlights the investigations on friction stir welding used to improve aluminum alloys, adhesion Phenomenon of liquid metals, secondary aluminum used for producing products, characteristic of an aluminum-silicon alloy in liquid state during the centrifugal casting process and review on melt pool convection during laser material processing.

Author details

Samson Jerold Samuel Chelladurai
Department of Mechanical Engineering, Sri Krishna College of Engineering and Technology, Coimbatore, Tamil Nadu, India

*Address all correspondence to: samsonjeroldsamuel@skcet.ac.in

Chapter 2

Investigations on Friction Stir Welding to Improve Aluminum Alloys

Bazani Shaik, Gosala Harinath Gowd and Bandaru Durga Prasad

Abstract

Today is an era of metals including Aluminum alloys owing to a fundamental paradigm shift in research objectives. In addition to superior performance and lightweight criteria that are used to define the innovations of yore, scientists today are compelled to take into consideration the environment-friendliness of the new and novel materials being developed due to the concerns of maintaining a sustainable and safe existence. The solid-state Friction stir welding process has immense potential in the areas of automobiles, aerospace and construction industries due to its overwhelming advantages over the conventional fusion welding process of aluminum alloys. The thesis presents an experimental investigation of friction stir welding of dissimilar aluminum alloys AA7075T651 and AA6082T651. Mathematical modeling equations are developed to predict the tensile strength, impact strength, elongation, and micro-hardness of the dissimilar FSW joints AA7075T651 and AA6082T651. The process parameters are optimized for maximum tensile strength and hardness values. Post weld heat treatment is conducted and the metallurgical properties of the FS welded AA7075T651 and AA6082T651 are presented for different combinations of tool rotational speeds. Aluminum and its alloys are widely used in nonferrous alloys for many industrial applications. Aluminum exhibits a combination of an excellent mechanical strength with lightweight and thus it is steadily replacing steel in industrial applications where the strength to weight ratio plays a significant role. In conventional welding, the joining of aluminum is mainly associated with a high coefficient of thermal expansion, solidification shrinkage and dissolution of harmful gases in the molten metal during welding. The weld joints are also associated with segregation of secondary alloys and porosities which are detrimental to the joint qualities. Friction Stir Welding (FSW) and Friction Welding (FW) are the most popular emerging solid welding techniques in aircraft and shipbuilding industries. FSW is mainly used for the joining of metal plates and FW is mainly used for the joining of rods. Both techniques are suitable for high strength material having less weight. These techniques are environmentally friendly and easy to execute. Hence, the study of these techniques can contribute much to the field of green technology. This research work is dealt with the experimental and numerical investigations on FSW and FW of aluminum alloys.

Keywords: Dissimilar Aluminium Alloys, Friction Stir Welding, Process Parameters, Tool

1. Introduction

Some of the previous research works already done by the researchers have been discussed hereafter. **Landry Giraud** et al. [1] Studied the AA7020T651 and AA6060T6 on friction stir welding of dissimilar heat treatable aluminum alloys 7020-T651 and 6060-T6. An experimental analysis is presented based on results obtained from temperatures and efforts measurements in a range of advance speed from 300 mm.min − 1 to 1100 mm.min − 1 and rotational speed from 1000 rev. min − 1 to 2000 rev.min − 1. Dissimilar welding does not seem to induce a hotter side but efforts are very sensitive to process parameters.

Prakash Kumar Sahua et al. [2] investigated dissimilar friction stir welding between aluminum (Al) and copper (Cu) at tool rotation rate of 1200 rev/min, welding speed of 30 mm/min, 0.1 mm plunging depth and 1.5 mm offset towards Al alloy yield highest ultimate tensile strength of 126.0 MPa and yield strength of 119.3 MPa which constitute 95% and 100%, respectively, of the 1050 Al base metal. The highest compressive strength and the bending angle were 7.8 MPa and 65°, respectively, for the specimen with the highest tensile strength. The hardness at the Cu side of the nugget is higher than that at the Al side. In the case of experiment E8 which yield the highest tensile strength, the maximum hardness at the NZ was 176 HV and the average hardness at NZ was 60 HV. Line scanning indicated a mixed flow of Al/Cu materials throughout NUGGET ZONE. **I.A. Kartsonakis** et al. [3] The dissimilar joining of AA6083T3 and AA5083H111 alloys with Tic nanoparticles, multi-wall carbon nanotubes, and cerium molybdate containers reinforcement. AA5083-H111 rich areas in the WN as well as the intergranular corrosion in the AA6082-T6 rich areas in the WN area. Moreover, MoO4 -2 ions that come from the container shell, are adsorbed onto the surface of both resulting in their protection of chloride penetration. Finally, the corrosion process is probably further inhibited due to the formation of cerium oxide films on the cathodic sites of the WN area. **M.-N. Avettand-Fènoël** et al. [4] Studied microstructural and mechanical characterization of an AA2024 – pure Cu linear friction weld. The interface is covered by a discontinuous layer of intermetallic compounds. Al_2Cu, AlCu, Al_2Cu_3 and more particularly Al_4Cu_9 were detected. Two metastable phases, i.e. an Al_3Cu_2 compound and an out-of-equilibrium Al solid solution containing 13 at % Cu, were also identified at the interface and on the Al side, respectively. The weld presents a joint coefficient lower than 0.5 for the yield strength and close to 0.3 for the ultimate tensile strength, and its brittle fracture, initiated by the intermetallic compounds, occurs at the interfacial zone. Due to this drawback, some routes of improvement of the Linear Friction Welding process are finally proposed.

U. Donatus et al. [5] Investigated AA6082T6 and Aa5083-O the effectiveness of the anodic oxide layer formed by a novel processing technique (pre-sputter-deposition prior to anodizing) for the corrosion protection of friction stir welds of dissimilar aluminum alloys. It is important to reiterate that the corrosion attack reported in this study are in two categories: (i) the pre-anodizing attacks caused by the anodizing solution at preferentially susceptible regions such as regions of aligned active second phase particles and grain boundaries which were observed in the HAZ of the AA6082-T6 alloy; and (ii) the post anodizing attacks caused by the etching solution which were particularly observed on the AA5083-O alloy regions. The pre-anodizing attack was completely prevented whilst the post-anodizing attack was significantly minimized by prior sputter-deposition of 1 μm thick pure Al before anodizing.

D.G. Hattingh et al. [6] Studied the development of a semi-automated friction stir welding (FSW) technique for joining 38 mm nominal outer diameter (OD) tubes of 6082-T6 aluminum alloy with 3.5 mm nominal wall thickness.

The technique incorporates a retracting tool in order to avoid leaving a substantial hole in the joint line after extracting the tool at the end of the welding process. This is one of the very first applications of FSW to small diameter tubular geometries to be reported in the open literature and the technique is capable of producing small-scale production runs (circa 100) of welded tube specimens with consistent tensile and fatigue properties. The tensile strength of the extruded 6082-T6 tube was 303 MPa while the joint efficiency of the weld was 0.55 both for complete tube specimens and for micro tensile specimens. This compares well with values reported in the literature for 3 mm flat plate specimens of 6082 alloy.

G.K. Padhy et al. [7] Studied recrystallization fractions in the stirred zone of Al 6061-T6 friction stir welds, prepared with and without ultrasonic vibrations, were evaluated using recrystallization fraction maps. Based on the maps, it was suggested that the microstructure evolution can be described as different dislocation manipulation processes. It was observed that the superposition of the static load of FSW on residual ultrasonic softening induces subgrain formation. Subgrain formation was substantial at the center of the stirred zone where the ultrasonic impact was the maximum.

Sivaraman Thapliyal et al. [8] Studied the effect of normal load, sliding velocity, and surface temperature on the dry sliding wear behavior of friction stir processed C95500 nickel aluminum bronze (NAB) alloy. Adhesive wear behavior was studied using the pin on disk tribometer as per ASTM G99-04 standard, using two-level full factorial approaches. The friction stir processed (FSPed) surface under study showed a lower wear rate than as-cast alloy. The wear rate in both conditions, as-cast as well as FSPed, at high temperature was found lower than the one during ambient temperature condition. The wear model and operating wear mechanism were assessed through SEM study of the worn surface and wear debris.

Ho-Sung Lee et al. [9] Studied for the joining of AA2195T0 and T8 FSW butt jointed, it is shown that the temperature in the advancing side is higher than the retreating side which supports the results of micro-hardness profile at nugget zone. For AA2195-T0, The weld zone can exhibit higher micro-hardness than parent material due to grain recrystallization and the hardness profile inside of the nugget depends on the cooling rate. For AA2195-T8, low hardness around the nugget zone is related to the dissolution and coarsening of precipitates at HAZ in the advancing side and/or retreating side. It is shown that the effect of removing oxide is effective on relatively low rotating speed conditions of 300 and 400 rpm, while as little influence to specimens welded higher rotating speed in this study.

Sergey Malopheyev et al. [10] Studied a simple but effective approach for improvement of strength of friction-stir welded 6061-T6 aluminum alloys was elaborated. It involves friction-stir welding (FSW) at relatively high tool rotational speed and welding speed followed by standard post-weld aging. The selected combination of FSW parameters provides high welding temperature as well as the rapid cooling rate. This leads to the complete dissolution of strengthening precipitates in the stir zone but hinders their coarsening in the heat-affected zone.

2. Microstructural investigations on aluminum alloys of AA7075T651 and AA6082T651 on FSW by using three different tools

The microstructural investigations on aluminum alloys AA7075T651 and AA6082T651 are done on the De-Wintor Inverted Trinocular Metallurgical Microscope shown in **Figure 1** and specifications of metallurgical microscope shown in **Table 1** and typical macrostructure of dissimilar of Al Alloys FSW joints showing at different zones shown in below **Figure 2**.

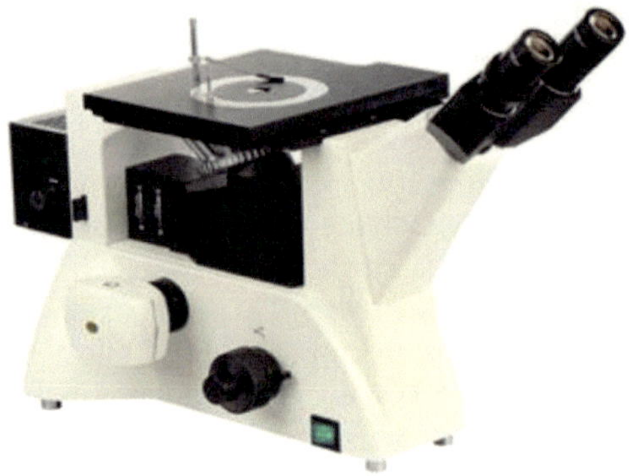

Figure 1.
De-Wintor inverted Trinocular metallurgical microscope.

SNo	Name	Specifications
1	Magnification	50X to 1000X
2	Eye Piece	Paired 15 X and 10 X
3	Objective	5 X, 10 X, 20 X, 25 X, 45 X, 50 X,100X
4	Power	12 Volt, 50 watts halogen lamp
5	Shaft	X-Y direction with 360^0 rotation
6	Polarization	With Polarizer prism
7	Microscale	Attached with 0.01 mm ocular scale

Table 1.
Specifications of metallurgical microscope.

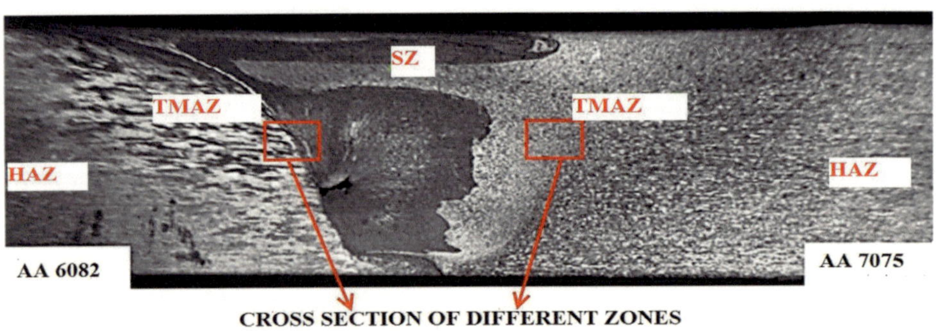

Figure 2.
Typical macrostructure of dissimilar Al alloys FSW joint showing different zones.

2.1 Microstructural investigations of AA7075T651 and AA6082T651 on FSW by using a square tool for samples

Figure 3 shows AA7075 of parent metal has magnification 100 x and etchant hydrofluoric solution are used its shows microstructure of AA7075 parent metal on the advancing side of the FSW process. The parent metal has an inrolled temper condition. The sheet has been cold worked by a rolling process, and primary grains of alpha aluminum is elongated along with direction forming. Eutectic constituents like Cu-Al_2, Mg_2Si, Zn-Al_2,and Mg-Al_2 precipitated with the direction of rolling.

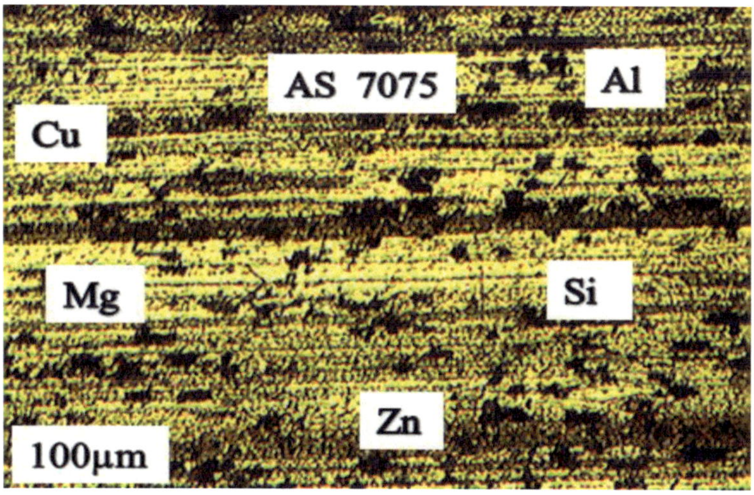

Figure 3.
AA7075 parent metal on advancing side on FSW by using a square tool for samples.

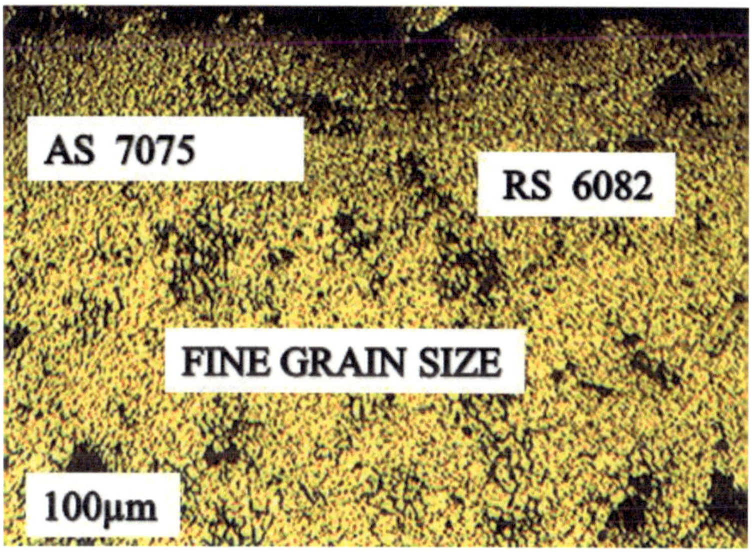

Figure 4.
Eutectic constituents of both AA7075 and AA6082 on FSW by using the square tool for samples.

Figure 4 shows the microstructures of advancing side AA7075 and retreating side AA6082 at shoulder zone of FSW process and the metal matrix underwent fragmentation facilitated the process dissolution of the eutectic constituents for both AA7075 and AA6082 due to frictional heat and stirring. The grains are finer and dynamic re-crystallization has formed as it is revealed by the fine size of the grains of primary aluminum.

Figure 5 shows AA7075 has a thermomechanical transformation zone where the heat process increases with the plasticity of the parent metal and metal has undergone plastic deformation in the direction of the tool.

Figure 6 shows the bottom zone of nugget with two distinct microstructures. The left side is parent metal AA7075 and the right side is nugget zone with fragmented constituents of both AA7075 and AA6082.

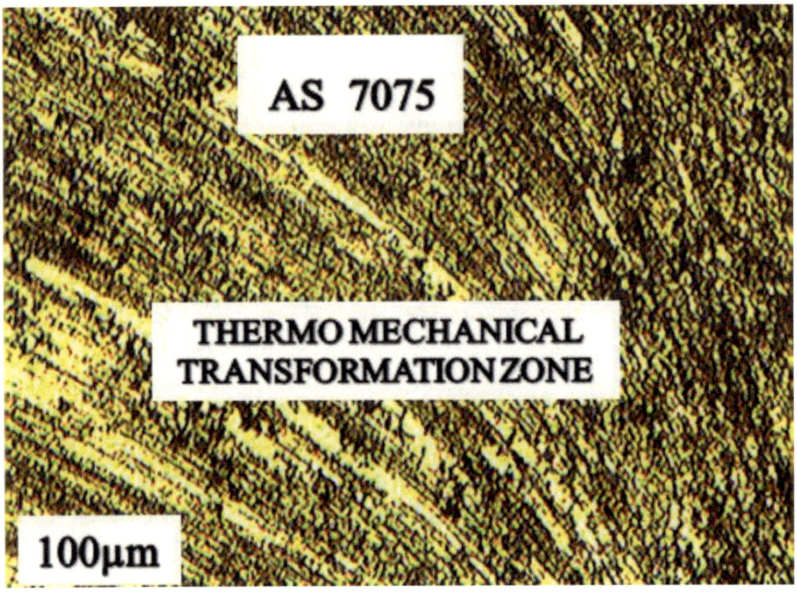

Figure 5.
TMT zone of AA7075 on FSW by using the square tool for samples.

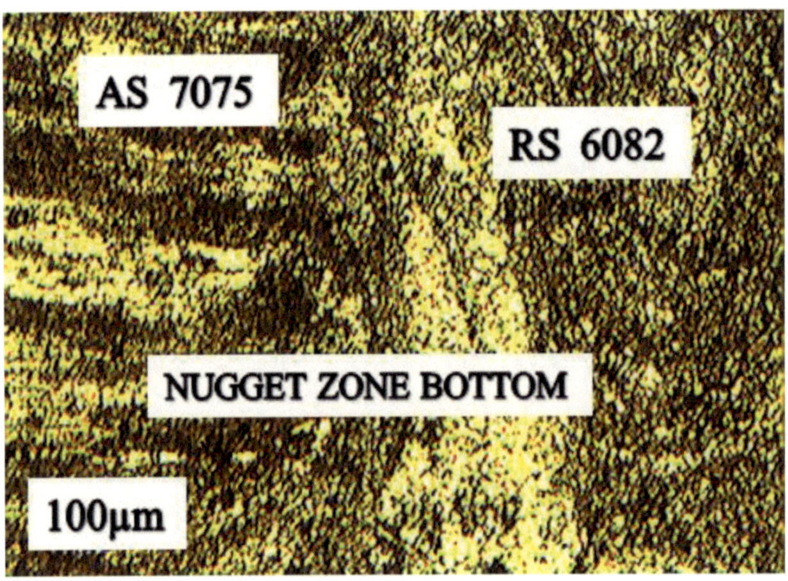

Figure 6.
Bottom zone of the nugget on FSW by using the square tool for samples.

Figure 7 shows microstructures of the upper zone of a nugget. The upper zone showed well re-crystallized grains and effective re-crystallization has taken place due to the conducive temperature existed. The grain size in **Figure 7** is 15 microns and with grain orientation towards the upper direction.

Figure 8 shows from the lower zone of the nugget zone. The grain size in **Figure 8** is mixed varying and elongated.

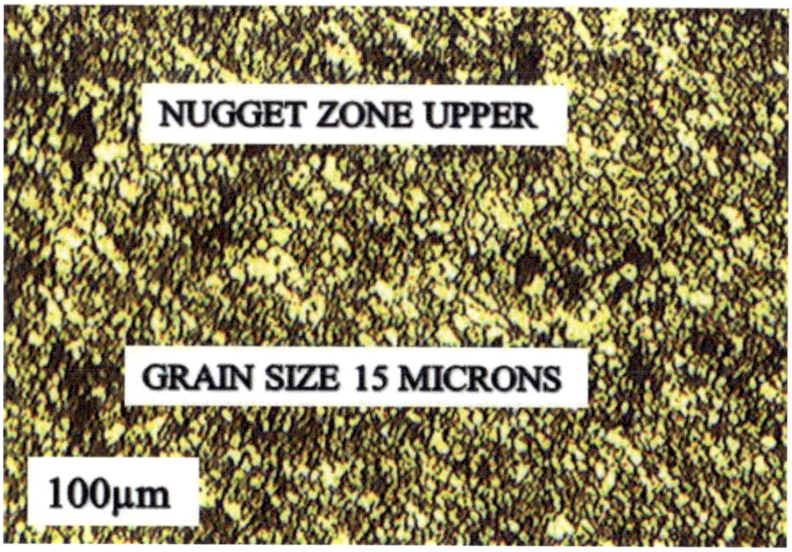

Figure 7.
The upper zone of the nugget on FSW by using the square tool for samples.

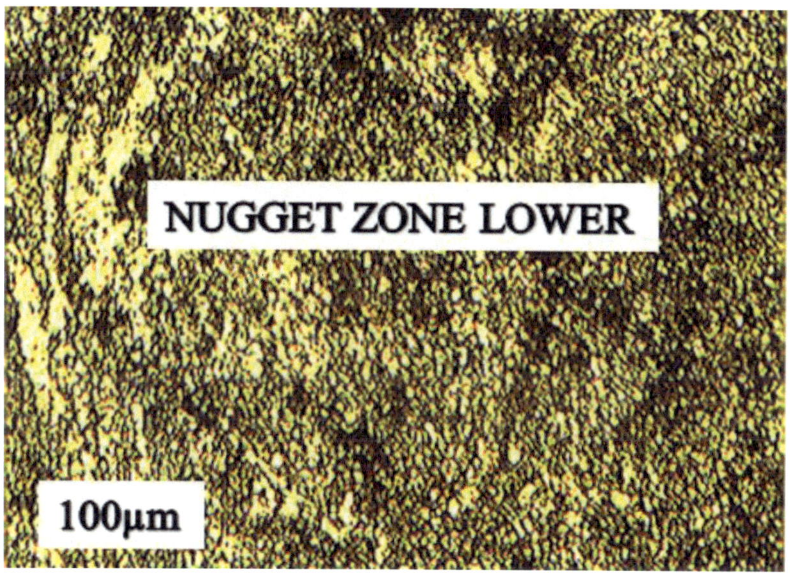

Figure 8.
The lower zone of the nugget zone on FSW by using the square tool for samples.

Figure 9 shows the interface zone of nugget and AA6082 on the right side and nugget on the left side.

Figure 10 shows parent metal AA6082 at the heat-affected zone with more precipitated particles of Mg_2Si in primary aluminum solid solution.

Figure 11 shows the microstructure of AA6082 has the orientation of grain size along the direction at the cold-rolled condition with rolling on eutectic and insoluble constituents.

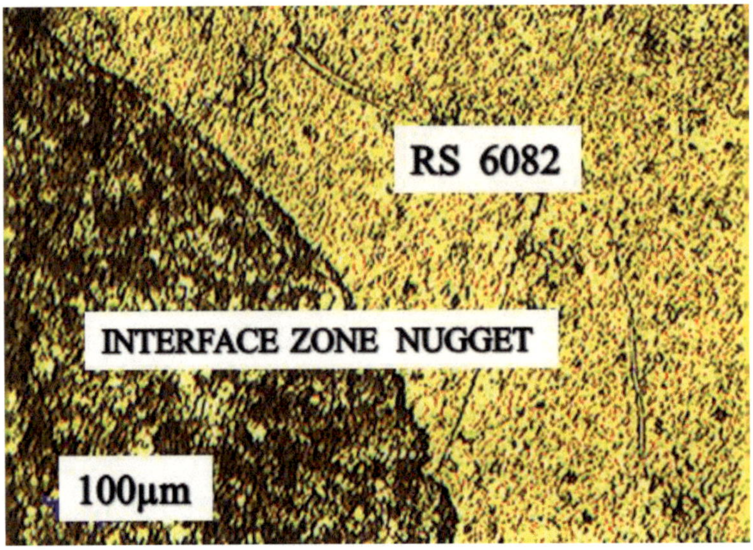

Figure 9.
Interface zone of the nugget on FSW by using the square tool for samples.

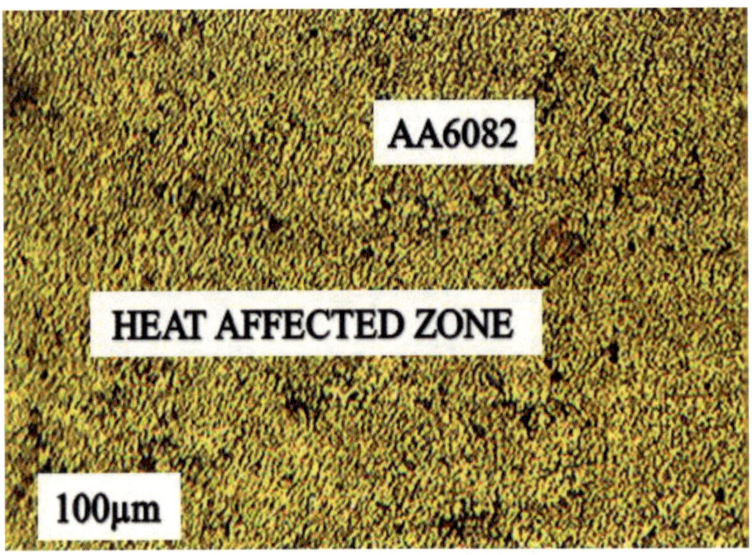

Figure 10.
Parent metal heat-affected zone of AA6082 on FSW by using a square tool for samples.

3. Conclusions

In this work, the important weld strength are analyzed after conducting experiments using the Friction Stir welding setup. The results presented in the work can be used for further analysis. That is using the experimental data empirical models can be developed and then these models can be used for finding the optimal process parameters to get the best output characteristics of welded joints. Then the problem can be solved by using an optimization algorithm after formulating the objective function. Later the entire process can be automated which helps to increase the production rate without increasing the unit cost of the welded joints.

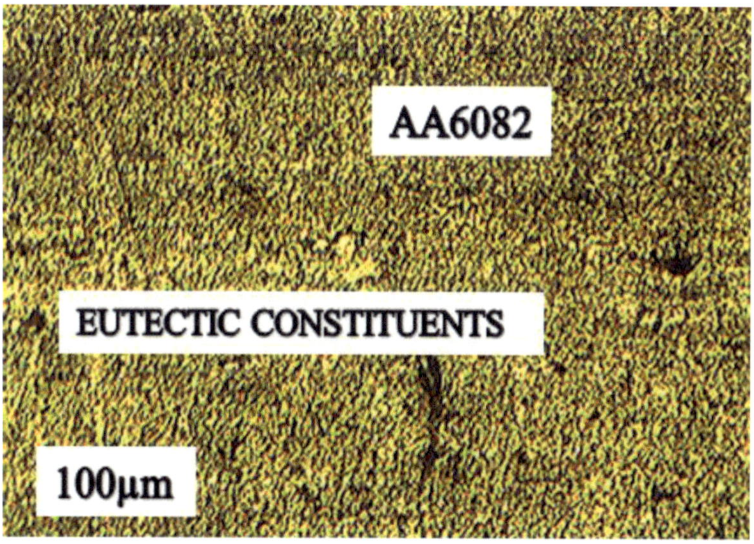

Figure 11.
AA6082 in cold rolled condition on FSW by using a square tool for samples.

For square tool in this work, the important weld strength. That is using the experimental data empirical models can be developed and then these models can be used for finding the optimal process parameters to get the best output characteristics of welded joints. Then the problem can be solved by using an optimization algorithm after formulating the objective function. Later the entire process can be automated which helps to increase the product rate without increasing the unit cost of the welded joints and microstructural changes of Nugget Zone Top and bottom, TMT zone of the AA 7075 T651 where the heat of the process increases the plasticity of the parent metal and metal has undergone plastic deformation in the direction of tool and eutectic constituents of Cu-Al, Mg-Al, Mg-Si of different grains formation are developed.

Author details

Bazani Shaik[1]*, Gosala Harinath Gowd[2] and Bandaru Durga Prasad[3]

1 Department of Mechanical Engineering, Ramachandra College of Engineering, Eluru, Andhra Pradesh, India

2 Department of Mechanical Engineering, Madanapalle Institute of Technology and Science, Madanapalle, Andhra Pradesh, India

3 Department Mechanical Engineering, Jawaharlal Nehru Technological University College of Engineering, Ananthapuramu, Andhra Pradesh, India

*Address all correspondence to: bazanijntua@gmail.com

References

[1] Landry Giraud, b, Hugo Robea, Christophe Claudina, Christophe Desrayaud, Philippe Bocher, Eric Feulvarcha," Investigation into the dissimilar friction stir welding of AA7020-T651 and AA6060-T6" Journal of Materials Processing Technology 235 (2016) 220-230-May 2016

[2] Prakash Kumar Sahu, Sukhomay Pal, Surjya K. Pal, Rahall Jain," Influence of plate position, tool offsets and tool rotational speed on mechanical properties and microstructures of dissimilar Al/Cu friction stir welding joints" Journal of Materials Processing Technology 235 (2016) 55-67)-April 2016.

[3] I.A. Kartsonakis, D.A. Dragatogiannis, E.P. Koumoulos, A. Karantonis, C.A. Charitidis" Corrosion behavior of dissimilar friction stir welded aluminum alloy reinforced with nano-additives, DOI: doi: 10.1016/j.matdes.2016.04.02, Reference: MADE 167, PII: S0264-1275(16)30492- -April 2016

[4] M.-N. Avettand-Fènoël, G. Racineux, L. Debeugny, R. Taillard "Microstructural characterization and mechanical performance of an AA2024 aluminum alloy — Pure copper joint obtained by linear friction welding, PII: S0264-1275(16)30311-2, DOI: doi: 10.1016/j.matdes.2016.03.02, Reference: JMADE 150- March 2016.

[5] U. Donatus, G.E. Thompson, X. Zhou "Effect of prior sputter deposition of pure aluminum on the corrosion behavior of anodized friction stir weld of dissimilar aluminum alloys" Scripta Materialia 123 (2016) 126-129.

[6] D.G. Hattingh, L.G. von Welligh D. Bernard L Susmel R. Tovo M.N. James "Semiautomatic Friction Stir Welding of 38 mm OD 6082-T6 Aluminium Tubes, Journal of Materials Processing Technology, DOI: http://dx.doi.org/doi:10.1016/j.jmatprotec.2016.07.027, PII: S0924-0136(16) 30253-9, Reference: PROTEC 14896-July 2016.

[7] G.K. Padhy, C.S. Wu and S. Ago, Sub grain formation in ultrasonic enhanced friction stir welding of aluminum alloy, Materials Letters, http://dx.doi.org/10.1016/j.matlet.2016.07.033-July 2016.

[8] Shivraman Thapliyal, Dheerendra Kumar Dwivedi "Study of the effect of friction stir processing of the sliding wear behavior of cast NiAl bronze: A statistical analysis" Teratology International 97 (2016) 124-135-January 2016.

[9] Ho-Sung Lee, Jong-Hoon Yoona, Joon-Tae Yoga and Kookie Nob" Friction Stir Welding Process of Aluminum-Lithium Alloy 2195" Procedia Engineering 149 (2016) 62 –June 2016, Novy Smokovec, Slovakia.

[10] Sergey Malopheyev, Igor Vysotskiy, Vladislav Kulitskiy, Sergey Mironov, Rustam Kaibyshev "Optimization of processing-microstructure-properties relationship in friction-stir welded 6061-T6 aluminum alloy" Materials Science & Engineering A 662 (2016) 136-143- March 2016.

Chapter 3

Progress on Experimental Study of Melt Pool Flow Dynamics in Laser Material Processing

Xianfeng Xiao, Cong Lu, Yanshu Fu, Xiaojun Ye and Lijun Song

Abstract

Laser material processing has becoming a rapid developing technology due to the flexibility of laser tool. Melt pool is the main product from the interaction between laser and material and its features has a great impact on the heat transfer, solidification behavior, and defects formation. Thus, understanding changes to melt pool flow is essential to obtain good fabricated product. This chapter presents a review of the experimental studies on melt pool flow dynamics for laser welding and laser additive manufacturing. The mechanisms of melt pool convection and its principal affecting factors are first presented. Researches on melt flow visualization using direct and indirect experimental methods are then reviewed and discussed.

Keywords: Laser welding, Laser additive manufacturing, Melt pool, Convection flow

1. Introduction

Replace the entirety of this text with the introduction to your chapter. The introduction section should provide a context for your manuscript and should be numbered as first heading. When preparing the introduction, please bear in mind that some readers will not be experts in your field of research.

Since laser was invented by Maiman in 1960, it has experienced rapid applications in laser material processing. The advantages of high quality, high precision, high efficiency and high flexibility promote laser welding and laser additive manufacturing becoming the best developing foreground technologies in welding areas and additive manufacturing domains, respectively. Unlike arc welding, laser welding creates small melt pool with a high intensity laser beam spot, which allows the achievement of smooth welding seam with narrow heat affected zone (HAZ) and low distortion. The noncontact feature of laser also frees the welder from harshest environments. Laser additive manufacturing's equipment and parameters share many common features with laser welding. Laser additive manufacturing can be considered by extending laser welding from two-dimension seam to three-dimension bulk with a synchronous powder or wire feeding. According to the ASTM F42 Committee [1], the laser powder bed fusion (LPBF) and laser directed energy deposition (L-DED) are the two most relevant laser additive manufacturing process.

The essence of laser material processing is laser interact with materials, either heat or melt, or ablation. Irradiated by a laser beam, the localized material undergoes a rapid heating–cooling thermal cycle. Melt pool is the main product from the interaction between laser and materials, both for laser welding and laser additive manufacturing. Melt pool temperature field and its evolution determine the temperature gradient (G) and solidification rate (R). Together, G and R determine the solidification morphology and the microstructure scale [2]. Namely, GR determines the scale of microstructure, their ratio G/R is linked to the morphology of solidified microstructure, thereby affecting the mechanical properties of the weld seam or fabricated part. In addition, melt pool geometry including its size and shape also affects solidification behavior. A wide and shallow melt pool beneficial to the epitaxial growth of grains along one direction, resulting in a strong texture. Melt pool dimension is also found correlated to residual stress in selective laser melting [3]. Such melt pool characteristics has been reviewed by Yan et al. [4], by Fotovvati et al. [5] and Willy et al. [6].

Another often overlooked melt pool characteristic is flow dynamics. Due to the small timescales and highly transient of the melt flow, it is very difficult to reveal dynamic behavior inside the melt pool. The flow in melt pool is mainly derived by spatial variation of the surface tension, which is known as thermocapillary flow or Marangoni convection, named after Italian Physicist Carlo Marangoni. Prior studies have shown that fluid flow plays an important role in heat transfer and solidification behavior in the melt pool, thus significantly affect the melt pool geometry, solidification microstructure, alloy element distribution, surface roughness and defects formation. Therefore, comprehensive understanding of the evolution of melt flow is a key concern and hot topic to improve the product quality during laser welding and laser additive manufacturing processes.

Melting of metals is commonly found various industrial applications, such as arc welding, metal casting and laser processing. Analysis of melting can be described as the Stefan problem assuming a heat conduction-controlled process; that is, fluid flow in molten pool is neglected. The effects of weld fluid flow induced by surface tension was first proposed in late 1960s [7]. From then on, various experimental and numerical investigations concerning fluid flow in laser melt pool have been reported in the open literature. Because of the small size of the weld pool and high dynamics, real-time experimental measurement of temperature and velocity fields inside the melt pool is very challenging. Therefore, mathematical modeling is the main research method to predict and describe the melt pool behavior. Mazumder [8] and Cook [9] have reviewed the approaches to incorporate melt convection effects as well as melt pool behavior during laser welding and laser additive manufacturing. However, quantitative investigations of flow pattern and velocity inside melt pool by experiment are still needed to validate the models. Recent developments in high-speed photography, image processing technology and the third-generation synchrotron radiation sources have enabled researchers to characterize the time-transient fluid flow inside the melt pool. Those research efforts are critical to reveal flow evolution and offers the possibility to calibrate or verify advanced numerical models.

This chapter aims to provide a comprehensive review of the experimental progress on melt pool flow dynamics for laser material processing, focusing on laser welding and laser additive manufacturing. The formation mechanism and driving forces for laser melt fluid flow is firstly introduced. Principal affecting factors for melt fluid flow are analyzed from open literature. The experimental results of laser melt fluid flow are reviewed and discussed, aiming at providing a fundamental understanding of melt flow convection mechanisms.

2. Forces within the melt pool and its effect factors

2.1 Driving forces in melt pool

The melt pool flow dynamics depends on the forces acting on the melt volume, thus the force analysis is crucial to investigate the formation mechanism of weld geometry, solidification microstructure, surface roughness and defects. The schematic diagram of forces on melt pool is shown in **Figure 1**. In the liquid melt pool during laser welding and laser additive manufacturing, there are four principal forces acting on the fluid flow: buoyancy force (originate from the spatial variation of the liquid-metal density), Marangoni force (originate from surface tension gradients), gravity and shear force (originate from laser induced vapor or plasma). In the case of applying an auxiliary electric or magnetic field, electromagnetic force on liquid melt pool should be also considered. Moreover, when evaporation occurs in the keyhole melting mode, recoil pressure becomes the principal driving force of molten metal. These driving forces and the interplay between them induce the complex flow motion in the melt pool.

Buoyancy force originate from the spatial variation of the liquid-metal density, mainly because of temperature variations, and, to a lesser extent, from local composition variations. It is known that density is a function of material's temperature, namely density decreases with increasing temperature. The temperature of liquid metal in the upper of the melt pool is higher than the bottom of the melt pool, leading to an upward movement of the melt pool as shown in **Figure 2**. Experimental [10] and numerical model [11] results have shown that buoyancy effect can be negligible when compared to Marangoni force in laser melting. The convection flow

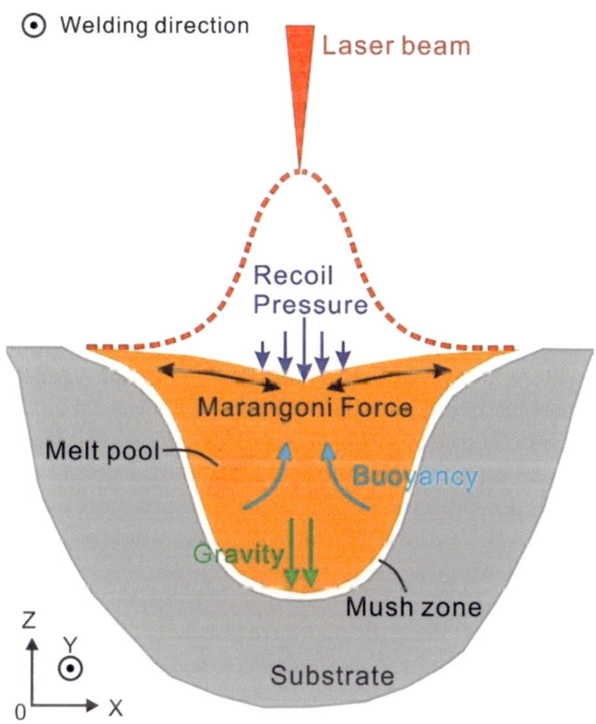

Figure 1.
Fluid forces acting on the weld pool.

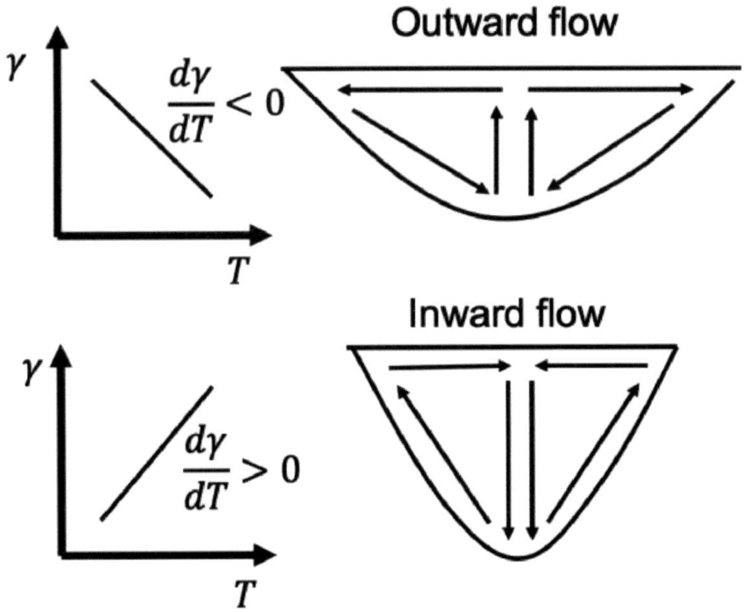

Figure 2.
Effect of the sign of the surface tension temperature coefficient on fluid flow in the weld pool.

caused by gravity is in the direction against the buoyancy force. Simulation results showed that gravity has no noticeable influence on the dimensions and shapes weld pool when laser welding a flat plate. However, when welding applied in circumferential condition or horizontal condition or near vacuum condition. The influence of gravity on the melt flow plays a critical role. The orientations of weld pools relative to gravity are different for different welding positions. For flat welding, gravity only contributes to the fluid flow in plate thickness direction. For inclined or horizontal welding, the melt pool is shifted afterwards under gravity action. Poor weld formation quality (unstable, porosity and undercut) is more likely to be developed [12]. Guo et al. demonstrated that full penetration of thick plate in horizontal position can mitigate some of the common welding defects including undercut and sagging [13].

One of the important aspects of laser welding and laser additive manufacturing is the convection driven by Marangoni force, also known as thermocapillary. The Marangoni force acts as a shear stress at the free surface thereby inducing convective flow within the molten pool. The driving mechanisms of surface tension can generally be classified as: temperature gradient, concentration gradient, pH gradient, surfactant-induced flow, and so on. Among these origins, temperature gradient is considered as the main driving forces of fluid flow in laser melting pure metals and most alloys. When laser locally heats the plate surface, the highest temperature located in the center of melt pool and decreases radially, causing a surface tension difference in the melt pool and thus creating an outward melt flow. A considerable amount of studies have showed the dominance of Marangoni force in the conduction mode melt pool convection.

Generally, two laser melting mechanisms: the conduction and the keyhole (deep penetration) mode are used. Recently, they are also adapted in laser additive manufacturing. Qualitative distinction of conduction mode and keyhole mode is whether evaporation happens or not. Once evaporation takes place, the vapor pressure (recoil pressure) acts like a piston on the liquid melt pool. The recoil pressure tends to push the liquid towards the pool edge and keyhole forms. Recoil pressure is widely accepted to be the principal driving force for fluid flow in the keyhole melting.

Besides the above-mentioned driving forces in melt pool, external force could be also introduced. For example, shielding gas in laser welding could help reduce surface oxidation and stabilize the melt pool fluctuations. It will also exert pressure on melt pool and alter the flow pattern in the molten pool. Electromagnetic force may be introduced via applying an electromagnetic compound field to the molten pool.

2.2 Factors affecting melt pool convection

According to the above force analysis in melt pool, surface tension and recoil pressure are the dominate driving mechanisms for melt pool convection. Processing parameters of laser welding and laser additive manufacturing can be classified into four types: laser related parameters, scan related parameters, gas/powder related parameters, material-related parameters.

2.2.1 Laser related parameters

Laser energy density is considered as one of the most significant variables on temperature field and material evaporation. Laser power, beam spot size, pulse frequency and energy distribution jointly determine the laser energy density. In conduction mode, the higher laser power leads to the larger temperature gradient in melt pool, resulting in higher surface tension and more intense radially convection. With the laser power increased, the input laser energy increased which caused an intense evaporation and the keyhole forms. Therefore, recoil pressure takes over as the primary driving force, pushing melt flow along the thickness direction.

2.2.2 Scan related parameters

For stationary laser welding, laser induced temperature field is axisymmetric resulting in an axisymmetric weld pool and keyhole. When laser moves with a certain velocity, temperature gradient in the front side of the moving laser beam is much steeper than that in the rear side. The melt pool shape resembles as in comet tail profile. Reducing laser scanning speed will cause the interaction time and peak temperature to increase substantially. As a result, increased temperature gradients lead to stronger Marangoni fluid convection and larger area of the molten pool. Laser oscillating welding is founded stabilize the fluid flow in melt pool and keyhole [14]. As for laser additive manufacturing of 3D bulks, scanning pattern, hatch spacing and layer thickness influence melt pool behavior through fore layers.

2.2.3 Gas/powder related parameters

In laser welding, side shielding gas serves three purposes: prevent the weld from oxidizing, remove the plasma plume and stabilize melt pool and keyhole. However, the too large flow rate of shielding gas gives resultant strong pressure on the melt pool and increases the fluctuation of the weld pool, keyhole and plasma. Thus, an optimal gas flow rate of shielding gas for a stable welding process is needed. For laser additive manufacturing, the main function of shielding gas is preventing melt pool from oxidizing. In DED process, typical average particle velocity is on the order of 5–10 m/s. The blown powder particles with low temperature impinging on the melt pool will change melt pool temperature field. In addition, impact force of powder particles may affect both the flow pattern and penetration of melt pool.

2.2.4 Material-related parameters

As mentioned previously, the melt flow in the pool is driven by surface tension gradients due to temperature gradients. The direction of the Marangoni flow is dictated by the sign of the surface tension gradient and is shown in **Figure 2**. For pure metals the surface tension coefficient is constant negative, therefore creating an outward radial flow (see **Figure 2(a)**). In 1982, Heiple and Ropper [15] found that the presence of surfactants in arc welding molten materials can alter Marangoni convection in the melt pool, and thus creating an inward radial flow (see **Figure 2(b)**). They also proposed that the Marangoni convection is the most important factor in determining weld shape, but without quantitative description of surface tension phenomenon. Sahoo et al. [16] were the first to propose a semi-empirical relationship between the surface tension gradient, temperature and content of surface-active elements, for various binary alloys. For a Fe-S binary alloy, at a certain of sulfur content, a critical temperature exists which corresponds to a change in the sign of the surface tension gradient, and results in a flow reversal, creating simultaneously two different vortices. Surfactant elements such as S, Se, Te, O can be added in the form of substrate, gas, wire or powder.

3. Experimental studies for melt pool flow investigation

Since surface tension driven weld fluid flow was first reported in 1965 [7], a number of experiment have been conducted to investigate the melt pool flow in laser welding and laser additive manufacturing. Melt pool flow investigation can be classified into indirect and direct approaches. Indirect methods by means of postmortem analysis of the cross sections of fusion zones are often used to infer the melt flow patterns. According to the employed equipment, direct observation of melt flow can be divided into three stages (see **Figure 3**): (1) 1970s ~ 2000s, use simulated material to visualize melt flow pattern; (2) 2000s ~ 2015s, employ high-speed camera and X-ray tube transmission system; (2) 2015s ~ to date, apply the third-generation synchrotron radiation sources for in-situ high-speed high-energy x-ray imaging.

3.1 Indirect analysis of the fluid flow

Melt flow convection in the weld pool will drive material transport in the weld. Indirect methods by means of postmortem analysis of the cross sections need tracers to identify the melt flow patterns. One way to analysis the melt flow pattern is using tracing particles with high hardness such as W, ZrO_2, SiC, TiB_2. Schemed as **Figure 4**, tungsten particles are pre-paved on the substate and two tungsten

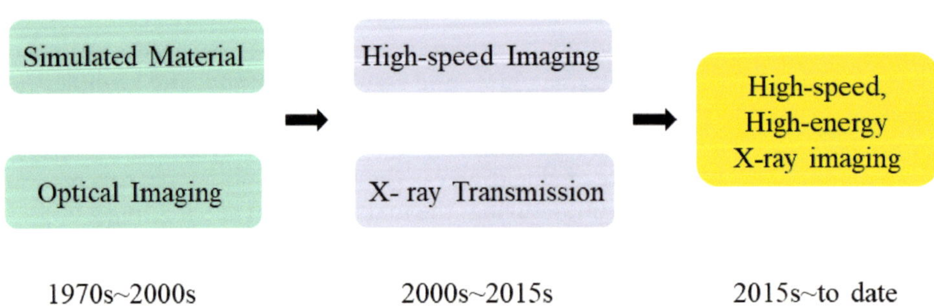

Figure 3.
Development of the experimental studies for direct observation of melt pool flow.

plates are inserted into the substate. After welding, microhardness distribution in the cross sections is measured. In the case of outward melt flow (**Figure 4(a)**), microhardness outside the plates is higher than that of between the plates. While for inward melt flow (**Figure 4(a)**), microhardness between the plates is higher than that of outside the plates. Li et al. [17] used this method to investigate the effect of shielding gas on TIG welding melt flow. Due to the high aspect ratio of welds in laser welding, it is difficult to insert two tungsten plates inside the substrate.

In 2005, Thomy and Vollertsen from BIAS [18] introduced a sandwich structure with a thin copper sheet between two aluminum sheets to study effects of magnetic fields on laser melt flow, see **Figure 5(a)**. With the help optical microscopy, hardness tester and EDX, darker region in **Figure 5(b)** and **(c)** is confirmed with higher copper content. Thus, the authors draw a conclusion that magnetic stirring induced by alternating magnetic fields promotes welds homogeneity. Beside using particles or metal sheet (Cu or Ni) as tracers, other forms of tracers such as filler has also been used [19].

3.2 Simulated materials for direct flow visualization

At the beginning, paraffin wax was employed to understand Marangoni flow in gas tungsten arc welding (GTAW). Ishizaki et al. [7] used a soldering iron to locally heat the surface of a thin slice of molten paraffin, circulation in the pool was observed by monitoring the movement of graphite particles. The resulting solidified structure that had a cross-sectional morphology similar to that of static GTAW welds. Similar Marangoni flow phenomena has also been found by using mercury [20], stearic acid [21], ice or water [22].

Simulated materials used above were applied in arc welding and only surface flow was visualized. For laser welding, it has a smaller size of melt pool and electromagnetic force does not exist. The physically simulated laser weld pool was first investigated by Limmaneevichitr and Kou from University of Wisconsin [23, 24]. Sodium nitrate, NaNO3, was chosen as it is transparent and exhibited similar surface properties to those seen in metal welding. In their experiment, a defocused CO_2 heat

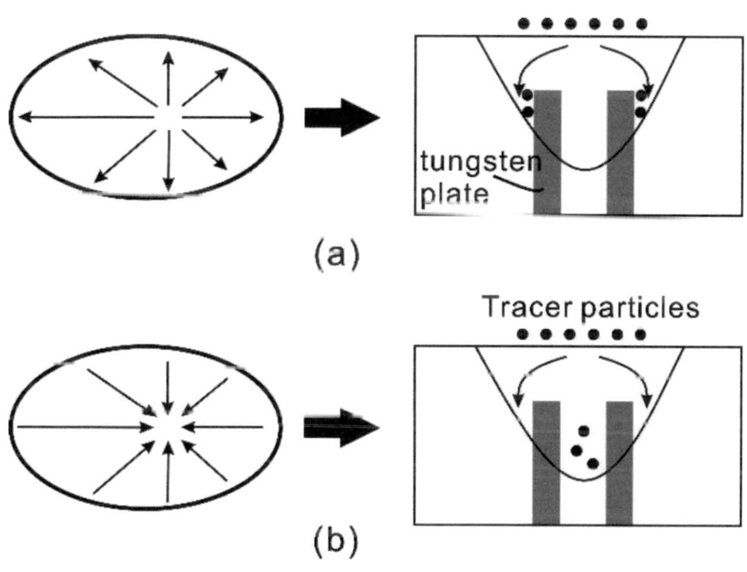

Figure 4.
Tracing particle distribution under (a) outward and (b) inward of Marangoni convection.

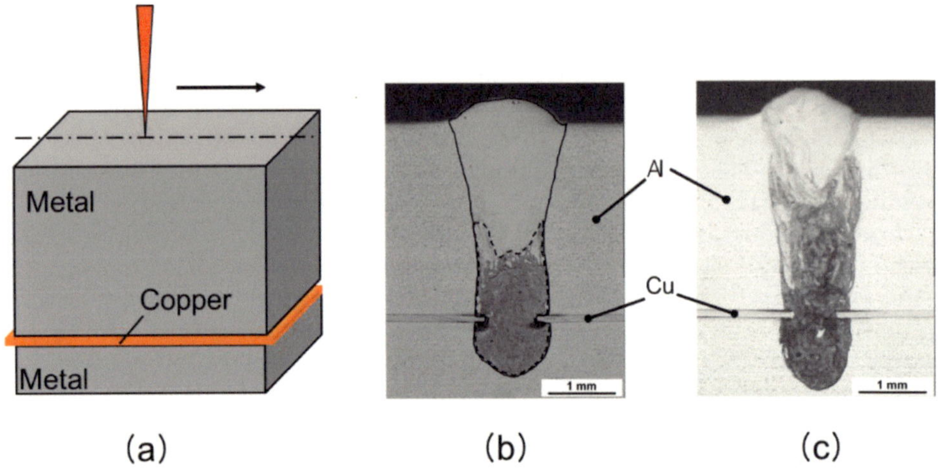

Figure 5.
(a) Scheme of sandwich structure with a thin copper sheet between two aluminum sheets; weld cross section with (b) omT and (c) 60 mT of alternating magnetic field [18].

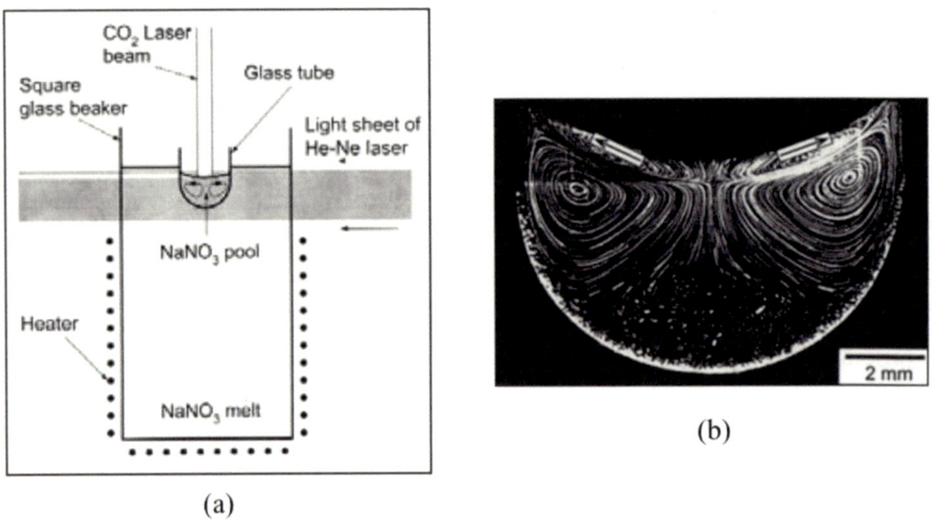

Figure 6.
(a) Experimental set up and (b) Visualization of Marangoni flows in a laser generated pool in a vat of $NaNO_3$ [23].

up the NaNO3 and another He-Ne laser light sheet, either vertical or horizontal, to cut through the pool to illuminate the tracer particles suspended in the pool and reveal the flow pattern, show in **Figure 6**. Also, to visualize the reversal flow pattern inside the weld pool, they used a transparent pool of NaNO3 with C_2H_5COOK as the surface-active agent [24]. This finding proved what was proposed by Heiple and Roper [15] in arc welding: a minor presence of surface-active elements can substantially change the temperature dependence of surface tension, leading to a change in flow pattern.

Another important driving force for melt flow is recoil pressure in the case of keyhole mode welding. Keyhole phenomenon is more complex and transient, due to fierce evaporation. The keyhole mode welding can be thermodynamically unstable and causes the formation of defects such as porosity, spatter, hump and undercut. Although direct observation of the keyhole is not easy, many efforts have been made. A low cost method for direct observation of keyhole and its evolution is welding transparent glass. Overall, there are three typical configurations with glass for keyhole observation: (1) directly welding on glass; (2) using sandwich structure consisting a metal foil between two glass plates; (3) using metal/glass structure. Schematic of the three configurations is shown in **Figure 7**.

The first image of keyhole was captured by Arata et al. in 1976, by welding in a soda-lime glass (**Figure 8(a)**). A similar approach was used on GG17 glass, which with an excellent heat-resistance property [25]. Sandwich structure consist of a metal foil between two glass plates (see **Figure 8(b)**) was first reported by Kato et al. [26] in 1985 in laser drilling. Zhang et al. used the sandwich structure approach to measure plasma inside keyhole [27]. It's worth noting that welding in glass (**Figure 8(a)**) and sandwich structure (**Figure 8(b)**) is far different from actual laser welding of compact metal. For welding in glass, physical and thermal properties of glass differ greatly from metallic engineering materials. For welding in sandwich structure, the loose multilayer construction is liable to cause keyhole collapse, which affects the stability of the welding process and could lead to misleading results. In 1994, Semak et al. [28] introduced a laser welding metal-glass approach to obtain transient keyhole profile, see **Figure 8(c)**. The penetration depth in real laser welding metal is used to calibrate the position of the laser beam center relative to the metal-glass interface. Suffering from limited high-speed imaging resolution, only low contrast keyhole profile was obtained. Nowadays, the rapid development of the high-speed imaging technology made it possible to observe the highly transient keyhole clearly. Zhang et al. [29] used a metal-glass samples which consists of one sheet of stainless steel and one piece of GG17 glass to directly observe the deep penetration welding keyhole, see **Figure 8(c)**. With the help of a high speed camera, a clear image of the keyhole wall was captured, shown as **Figure 8**. In recent years, the metal/glass structure approach is widely used to study keyhole dynamics and welding defects formation mechanisms [30, 31].

3.3 High speed imaging of tracers

It is worth noting that both the element tracing method and simulated material method can only drive quantitative conclusions. Addition, element tracing method can only obtain the final state of melt flow, lacking of transient information. Laser

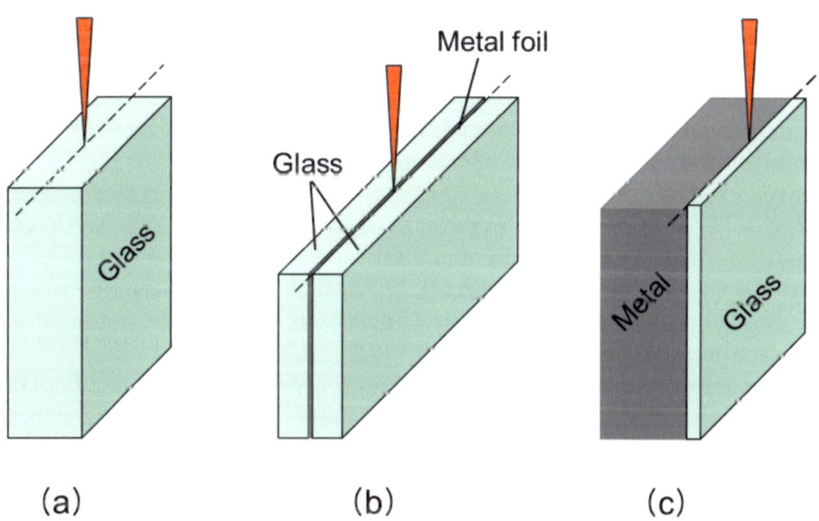

Figure 7.
Three typical configurations of keyhole observation for laser welding in: (a) glass, (b) glass/foil/glass, (c) metal/glass.

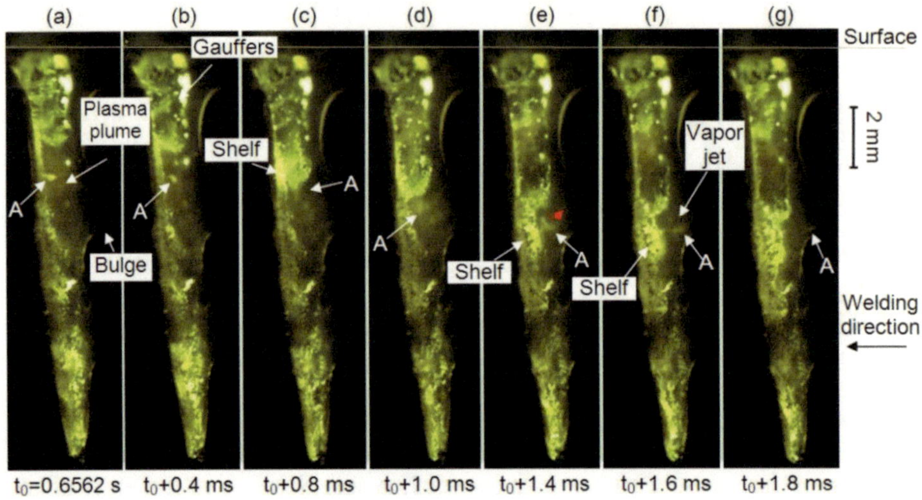

Figure 8.
Clearly keyhole images captured by laser welding metal/glass structure, (a)-(g) with a interframe time of 0.4 ms [29].

weld pool flow dynamics have been studied by simulation for many years since it is difficult to visualize transient flows in such a tiny zone. With the rapid development of high speed camera equipment and imaging processing technology, researchers can now capture highly transient melt flow.

Since melt flow velocity in laser melting can be of the order of 1 m/s [32, 33], successive images of a single tracer should be captured in a time interval shorter than 1 ms, corresponds to a minimum frame rate of 1000 fps. Therefore, imaging frame rate must be kept in the multi-fps range. In order to capture higher resolution and large viewing areas of melt pool, external illumination with narrow band interference filter is necessary. This technique reduces greatly the effect of laser induced vapor or plasma radiation.

To quantify the melt flow velocity, tracer-based flow measurement methods are wildly used. The tracer could either be “nature”, that is to say belong to the weld pool (such as surface oxide particles, humps, slag particles), or ‘artificial’ particles, added by the experimenter. Calculation of melt flow velocity is by measuring the distance between the tracer in two successive images divided by the interval time. For low density of tracers, particle tracking velocimetry (PTV) algorithm is applicable. While for moderate density of tracers, particle image velocimetry (PIV) is widely been used by tracking groups of tracers and performing a cross-correlation calculation on successive images. In the case of a very high density of tracers, optical flow type of approach is needed. Ki et al. [32] use a hump as the tracer to measure the melt flow velocity, shown as **Figure 9**. The authors assumed that the velocity of hump is close to the actual flow velocity. Thus, the experimentally obtained weld melt flow velocity were in the range of 1.4 to 2.2 m/s.

However, PTV-based method can only obtain several path lines in a flow field. In order to get the whole picture of melt flows, more “nature” or ‘artificial’ tracers are needed. Wirth et al. [33] used particles tacking method to obtain melt pool surface flow field. The results shown that using metal powder particles as tracers has a qualitatively similar flow field with using carbide or oxide particles as tracers, shown as **Figure 10**. The flow lines arisen from the center of melt poot point to its edge indicates that melt flow is driven by Marangoni force caused by temperature gradient in L-DED.

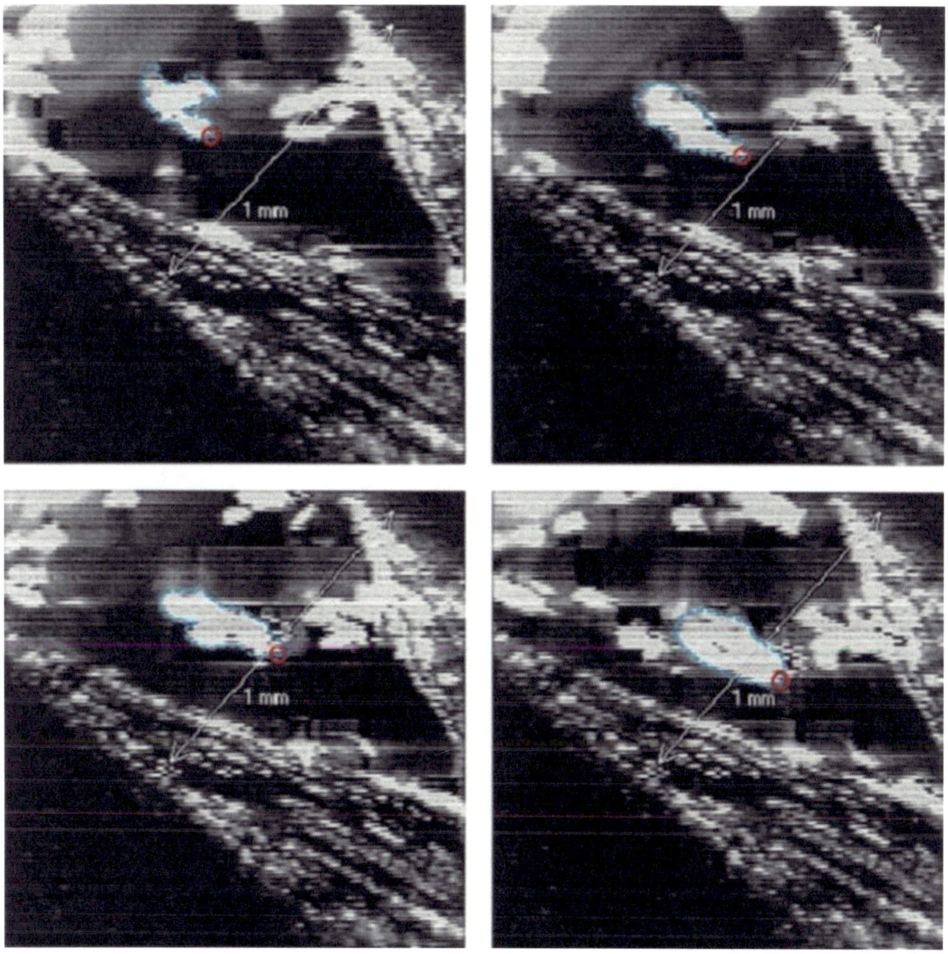

Figure 9.
Four successive images of a hump tip during laser welding [32].

3.4 In suit X-ray imaging

Besides efforts on studying melt surface flow dynamics, internal flows have also been the subject of intensive researches by many researchers. The in situ X-ray transmission imaging technique is a very useful tool to visualize the invisible phenomena in the laser melting sample. The first reported work on X-ray transmission imaging of welding dynamics was in electron beam welding by Arata et al. in 1976 [34]. Later, intensive investigations on keyhole formation [35], keyhole collapse [36], and keyhole porosity formation [37] have been conducted by the laser group in Osaka University around 2000s. **Figure 11** shown the X-ray transmission imaged keyhole melt flow by tracing tungsten particle and porosity formation during laser welding [38]. A more advanced X-ray transmission imaging system based on X-ray tube source was developed by Abt et al. [39] from IFSW, Germany.

Due to the low spectral intensity of X-ray tube, it is hard to observe clear solid–liquid interface. Recently, with the advent of high-flux, high-energy third-generation synchrotrons, X-ray phase contrast imaging is by far the most effective technique for revealing sub-surface structural dynamics with extremely high spatial and temporal resolutions. The knowledge gained are revealing new insights in laser welding and laser additive manufacturing. High-flux, high-energy synchrotron

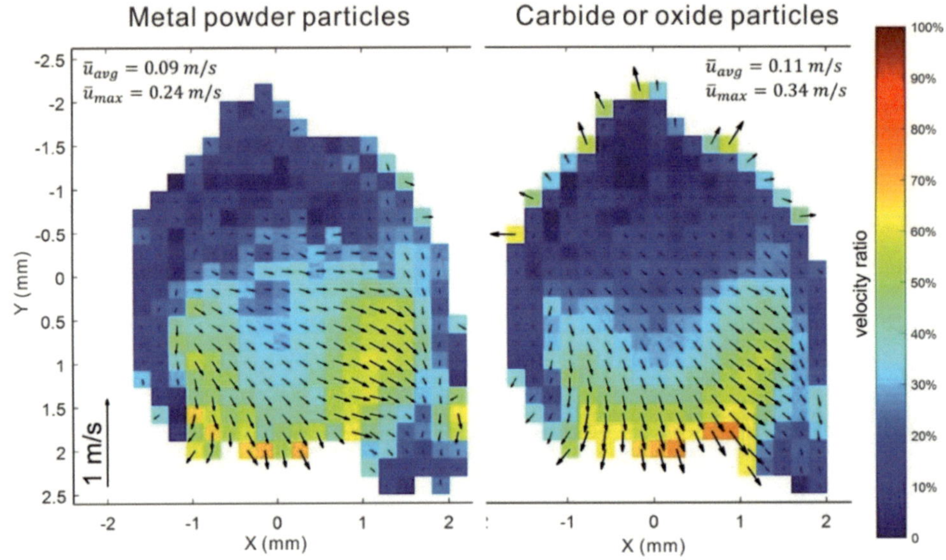

Figure 10.
Resulting metal powder particles (left) and carbide/oxide particles (right) velocity field during L-DED [33].

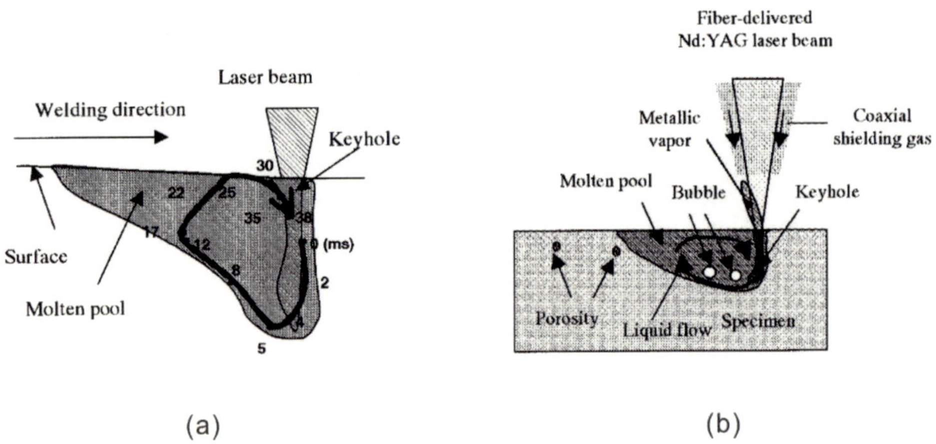

Figure 11.
(a) Keyhole melt flow and (b) porosity formation observed by in situ X-ray transmission imaging setup [38].

X-ray beam source	X-ray energy	Field of view (FOV) (width × height)	Spatial resolution	Time Resolution	References
Super Photon ring-8 GeV (Spring-8), Japan	3–70 keV	24 × 5.1 mm	38 μm	15 ms (70fps)	[40]
Diamond Light Sourcem (DIAMOND), UK	55 keV	8.4× 3.3 mm	6.6 μm	196 μs (5100 fps)	[41]
Stanford Synchrotron Radiation Lightsource (SSRL), USA	24 keV	2.2 × 2.2 mm	1.1 μm	250 us (4000 fps)	[42]
Advanced Photon Source (APS), USA	24.4 keV	1.5 × 1.5 mm	1 μm	100 ps (6500 kfps)	[43]

Table 1.
Several high-speed, high-energy synchrotron facilities used for in situ imaging laser melting process.

X-ray imaging, such as available at the Super Photon ring-8 GeV (Spring-8), Japan, the Diamond Light Source, UK, the Stanford Synchrotron Radiation Lightsource (SSRL), USA and the Advanced Photon Source (APS), USA, has been used to capture the keyhole behavior and defects formation clearly. **Table 1** summaries parameters of these facilities used for in situ imaging laser melting process.

4. Conclusion

In this chapter, we summarized the experimental studies on melt pool flow dynamic during laser material processing, focusing on laser welding and laser additive manufacturing. To visualize the melt pool flow patterns and velocity field, indirect and direct methods have been employed. Indirect methods are simple with low cost, but it can only achieve the final melt flow patterns by postmortem analysis of the cross sections of fusion zones. Direct methods include simulated materials, high speed imaging, and in situ X-ray transmission imaging. These three direct methods need experimental conditions from low cost to expensive, and reveal melt flow information from qualitative to quantify, from surface to internal. This chapter provides a generic guideline for experimental studying melt fluid flow dynamics.

Acknowledgements

The authors acknowledge the support of the National Natural Science Foundation of China (Grant No. 51875190 and 11662010), and Jiangxi National Natural Science Foundation of China (20192BCB23003).

Author details

Xianfeng Xiao[1], Cong Lu[1], Yanshu Fu[1], Xiaojun Ye[1] and Lijun Song[2]*

1 School of Mechatronics Engineering, Nanchang University, Nanchang, Jiangxi, China

2 State Key Laboratory of Advanced Design and Manufacturing for Vehicle Body, Hunan Provincial Key Laboratory of Intelligent Laser Manufacturing, Hunan University, Changsha, China

*Address all correspondence to: ljsong@hnu.edu.cn

References

[1] ASTM F42 Committee: Standard Terminology for Additive Manufacturing Technologies. ASTM International, 2012.

[2] Kou S. Welding Metallurgy. 2nd ed: John Wiley & Sons; 2003.

[3] Staub A, Spierings AB, Wegener K. Correlation of meltpool characteristics and residual stresses at high laser intensity for metal lpbf process. Advances in Materials and Processing Technologies. 2018:1-9.

[4] Yan Z, Liu W, Tang Z, Liu X, Zhang N, Li M, et al. Review on thermal analysis in laser-based additive manufacturing. Optics & Laser Technology. 2018;106:427-41.

[5] Fotovvati B, Wayne SF, Lewis G, Asadi E. A Review on Melt-Pool Characteristics in Laser Welding of Metals. Advances in Materials Science and Engineering. 2018;2018:1-18.

[6] Willy HJ, Li X, Tan YH, Chen Z, Cagirici M, Borayek R, et al. Overview of Finite Elements simulation of temperature profile to estimate properties of materials 3D-printed by Laser Powder-Bed Fusion. Chinese Physics B. 2020;29(4):048101.

[7] K Ishizaki NA, H Murai. Penetration in arc welding and convection in molten metal. Journal of Japan Welding Society. 1965;34:146-53.

[8] Mazumder J. Overview of melt dynamics in laser processing. Optical Engineering. 1991;30(8):1208.

[9] Cook PS, Murphy AB. Simulation of Melt Pool Behaviour during Additive Manufacturing: Underlying Physics and Progress. Additive Manufacturing. 2019;31:100909.

[10] Martin AA, Calta NP, Hammons JA, Khairallah SA, Nielsen MH, Shuttlesworth RM, et al. Ultrafast dynamics of laser-metal interactions in additive manufacturing alloys captured by in situ X-ray imaging. Materials Today Advances. 2019;1:100002.

[11] Srinivasan J, Basu B. A numerical study of thermocapillary flow in a rectangular cavity during laser melting. International Journal of Heat and Mass Transfer. 1986;29(4):563-72.

[12] Chang B, Yuan Z, Pu H, Li H, Cheng H, Du D, et al. A Comparative Study on the Laser Welding of Ti6Al4V Alloy Sheets in Flat and Horizontal Positions. Applied Sciences. 2017;7(4):376.

[13] Guo W, Liu Q, Francis JA, Crowther D, Thompson A, Liu Z, et al. Comparison of laser welds in thick section S700 high-strength steel manufactured in flat (1G) and horizontal (2G) positions. CIRP Annals. 2015;64(1):197-200.

[14] Zhang C, Li X, Gao M. Effects of circular oscillating beam on heat transfer and melt flow of laser melting pool. Journal of Materials Research and Technology. 2020;9(4):9271-82.

[15] CR Heiple HC. Mechanism for Minor Element Effect on GTA Fusion Zone Geometry. Welding Journal. 1982;61(4):97-102.

[16] Sahoo P, Debroy T, McNallan MJ. Surface tension of binary metal—surface active solute systems under conditions relevant to welding metallurgy. Metallurgical Transactions B. 1988;19(3):483-91.

[17] Li D, Lu S, Li D, Li Y. Tracer investigation of convection in weld pool under TIG welding process (in Chinese). Transactions of the China Welding Institution. 2011;32(8):45-8.

[18] Thomy C, Vollertsen F. Influence of Magnetic Fields on Dilution during Laser Welding of Aluminium. Advanced Materials Research. 2005;6-8:179-86.

[19] Zhao L, Sugino T, Arakane G, Tsukamoto S. Influence of welding parameters on distribution of wire feeding elements in CO2laser GMA hybrid welding. Science and Technology of Welding and Joining. 2009;14(5):457-67.

[20] Woods RA, Milner DR. Motion in the weld pool in arc welding. Welding Journal. 1971;50(4):163s-73s.

[21] Andersson D. Streaming due to a thermal surface tension gradient. Scientific Paper No TRITA-EPP-73-19, Royal Institute of Technology, Stockholm, Sweden. 1973.

[22] Ishizaki K. Interfacial tension theory of arc welding phenomena: formation of welding bead. Journal of Japan Welding Society. 1965;34(2):146.

[23] Limmaneevichitr C, Kou S. Visualization of Marangoni convection in simulated weld pools. Welding Journal. 2000;79:126-s.

[24] Limmaneevichitr C, Kou S. Visualization of Marangoni Convection Simulated Weld Pools Containing a Surface-Active Agent. Welding Journal. 2000;79:324-s.

[25] Jin X, Li L. An experimental study on the keyhole shapes in laser deep penetration welding. Optics and Lasers in Engineering. 2004;41(5):779-90.

[26] Kato J, Takamasu K, Ozono S. Investigation of material expulsion mechanism in laser drilling using modelled work piece. Bulletin of the Japan Society of Precision Engineering. 1985;19(2):133-4.

[27] Zhang Y, Li L, Zhang G. Spectroscopic measurements of plasma inside the keyhole in deep penetration laser welding. Journal of Physics D: Applied Physics. 2005;38(5):703.

[28] Semak VV, Hopkins JA, McCay MH, McCay TD. Dynamics of penetration depth during laser welding. Proceedings of ICALEO. 1994:G830-7.

[29] Zhang M, Chen G, Zhou Y, Li S. Direct observation of keyhole characteristics in deep penetration laser welding with a 10 kW fiber laser. Optics express. 2013;21(17):19997.

[30] Huang L, Liu P, Zhu S, Hua X, Dong S. Experimental research on formation mechanism of porosity in magnetic field assisted laser welding of steel. Journal of Manufacturing Processes. 2020;50:596-602.

[31] Ning J, Zhang L-J, Yin X-q, Zhang J-X, Na S-J. Mechanism study on the effects of power modulation on energy coupling efficiency in infrared laser welding of highly-reflective materials. Materials & Design. 2019;178:107871.

[32] Ki H, Mazumder J, Mohanty PS. Modeling of laser keyhole welding: Part II. simulation of keyhole evolution, velocity, temperature profile, and experimental verification. Metallurgical and Materials Transactions A. 2002;33(6):1831-42.

[33] Wirth F, Arpagaus S, Wegener K. Analysis of melt pool dynamics in laser cladding and direct metal deposition by automated high-speed camera image evaluation. Additive Manufacturing. 2018;21:369-82.

[34] Arata Y, Abe E, Fujisawa M. A study on dynamic behaviours of electron beam welding (report I): The observation by a fluoroscopic method. Transactions of Japan Welding Research Institute. 1976:1-9.

[35] Arata Y, Abe N, Oda T. Beam hole behaviour during laser beam welding. Proceedings of ICALEO. 1983;38:59-66.

[36] Matsunawa A, Kim J-D, Seto N, Mizutani M, Katayama S. Dynamics of keyhole and molten pool in laser welding. Journal of Laser Applications. 1998;10(6):247.

[37] Matsunawa A, Kim J, Katayama S. Porosity formation in laser welding mechanisms and suppression methods. Proceedings of ICALEO. 1997:G73-82.

[38] Katayama S, Kobayashi Y, Mizutani M, Matsunawa A. Effect of vacuum on penetration and defects in laser welding. Journal of Laser Applications. 2001;13(5):187.

[39] Abt F, Boley M, Weber R, Graf T, Popko G, Nau S. Novel X-ray System for in-situ Diagnostics of Laser Based Processes – First Experimental Results. Physics Procedia. 2011;12:761-70.

[40] Dorsch F, Terada T, Yamada T, Nishimura A. Development of laser cladding system with process monitoring by x-ray imaging. High-Power Laser Materials Processing: Lasers,Beam Delivery,Diagnostics,and Applications III. 2014;8963:1-10.

[41] Leung CLA, Marussi S, Atwood RC, Towrie M, Withers PJ, Lee PD. In situ X-ray imaging of defect and molten pool dynamics in laser additive manufacturing. Nature Communications. 2018;9(1).

[42] Calta NP, Wang J, Kiss AM, Martin AA, Depond PJ, Guss GM, et al. An instrument for in situ time-resolved X-ray imaging and diffraction of laser powder bed fusion additive manufacturing processes. Review of Scientific Instruments. 2018;89(5):055101.

[43] Parab ND, Zhao C, Cunningham R, Escano LI, Fezzaa K, Everhart W, et al. Ultrafast X-ray imaging of laser–metal additive manufacturing processes. Journal of Synchrotron Radiation. 2018;25(5):1467-77.

Section 2

Liquid Metals

Chapter 4

Adhesion Phenomenon of Liquid Metals

Hadef Zakaria and Kamli Kenza

Abstract

In this chapter, we study an interfacial phenomenon between liquid metals and ceramic substrates. Therefore, investigation of these phenomena is of great importance not only in technological applications but also in fundamental understanding of physical behavior of the adhesion between two different materials as far as their electrical structures and physiochemical properties are concerned. Moreover, adhesion energy is interpreted thermodynamically by the interfacial interactions and the nature of bonding between liquid metal and ceramic material. The adhesion energy in metal/ceramic systems is determined by using an electro-acoustical model based on the propagation of the acoustic wave in the interface and strongly depends on the electric properties of combination.

Keywords: Liquid metal, Ceramics, Adhesion, Interfaces, Gap energy, Acoustic parameters

1. Introduction

Metalized ceramics by liquid metal have a crucial uses in several modern technological applications such as solar cell [1–4] electrical devices [5–7] and Micro Electro Mechanical Systems (MEMS) [8–10]. Recently, these systems are used as the conductive wiring of microelectronic circuits; there has been considerable interest in the characterization of the structure and properties of liquid metal/ceramic interface [11].

However, the coating of ceramic surfaces can affect most of the properties of the interface. Therefore, the investigation of interfacial phenomena between metals and ceramic substrates is of great importance not only in technological applications but also in fundamental understanding of physical behavior of the adhesion between two different materials as far as their electrical structures and physiochemical properties are concerned. In fact, at the interface of a metal/ceramic system, adhesion occurs when the atoms or molecules of the two contacting surfaces approach each other so closely that attractive forces between approaching atoms (or molecules) bond them together. The strength of the bond depends on the size of the atoms, the distance between them, and the presence or absence of contaminant matter on the surface [1]. Hence, the strength or wea1kness of bonds is the key factor to determine the interface stability: good adhesion, welded adhesion, perfect bonding, weak bonding smooth interface, etc. The metal/ceramic contact is characterized by the adhesion energy, W_{ad}, which is the work per unit area of the interface needed to separate reversibly a metal/ceramic interface [2]. This physicochemical parameter is given by Young-Dupré equation relating surface tension of molten metal above melting temperature, γ_{LV}, and measured equilibrium contact angle θ formed between deposited liquid metal and its ceramic substrate [12]:

$$W_{ad} = \gamma_{LV}\,(1 + cos\theta) \tag{1}$$

Adhesion energy represent in generally the sum of all interfacial interactions between two surfaces [13]:

$$W_{ad} = W_{non-equil} + W_{equil} \tag{2}$$

$W_{non\text{-}equil}$ and W_{equil} represents non-equilibrium and equilibrium contributions respectively of interfacial interactions. The first term does not exist in the absence of chemical reactions, and the second term corresponds to non-reactive metals/ceramic systems [13], this later expressed by:

$$W_{equil} = W_{VDW} + W_{chem-equil} \tag{3}$$

W_{VDW} is van der Waals interaction and $W_{chem\text{-}equil}$ is chemical equilibrium interactions accompanied by formation of these chemical bonds between two contact phases. It is imported to note that these interfacial bonds rested without rupture contrary in non-equilibrium systems [13].Van der Waals energy in metal/ceramic systems estimate the can be numerically estimated by considering the dispersion interaction between a pair of atoms:

$$W_{VDW} = n\frac{3\alpha_M\alpha_C}{2R^6}\left[\frac{I_M I_C}{I_M + I_C}\right] \tag{4}$$

α_M and α_C are the polarizability volume of metal and ceramic; I_M and I_C the first ionization potential of metal atom and ceramic atom respectively. R is the distance between centers of the interacting atom.

At the interface zone, the surface acoustic wave (SAW) propagation which depends on elastic properties of solid substrates is greatly affected: the response would be different depending on the weakness or strength of bonds due to impedance mismatching [14]. Hence, in this context, we investigate the dependences of adhesion energy on acoustic parameters, in particular SAW velocities, for many metal/ceramic systems.

The objective of the electro-acoustic model [15] is the investigation of interfacial adhesion in liquid metal/ceramic systems subjected to non-reactive wetting in order to eliminate the non-equilibrium contribution Wnon-equil of adhesion work during a chemical reaction at the interface. A wide range of non-reactive liquid metals were used in this proposed model.

2. Choice of liquid metals

At the room temperature, most metals have a crystalline phase; the most widely used are iron, aluminum and copper. They are often present in oxide form (sodium oxide, magnesium oxide ...), some metals are present in the non-oxidized state (precious metals: platinum, gold) or in the form of alloys. Metal alloys are in general the combination of two or more metals as in the case of brasses (alloys of copper and zinc); but they can also contain non-metallic elements (i.e. iron-carbon alloy). Metals and their alloys are usually very good conductors of heat and electricity; they are most often hard, rigid and plastically deformable. It should be noted that a large number of metals have a very high melting point, since they have relatively weak mechanical properties and are most often characterized by a wettability, a low thermal and electrical conductivity (as in the case of copper and gold). Therefore, the use of metals in metallized ceramic structures requires a fusion process in order to liquefy or melt these metals. For this, the role of metallization is to make the ceramic wettable by the liquid metal.

Several liquid metals parameters used in this investigation are listed in **Table 1**; sound velocities at melting temperatures are tabulated by Blairs [16], surface tension values are proposed by Keene [17], Liquid densities are taken by Crawley [18] and by Blairs [16]. Whiles the elastic constants, solid densities and Rayleigh velocity are obtained from Briggs [19].

Metal	c (m/s)	γ_{LV} (mJ/m^2)	P_{lm} (Kg/m^3)	T_f (K)	E (GPa)	ρ_{sm} (Kg/m^3)	V_{RM} (m/s)
Na	2526	203	951	371	10	968	1875
Mg	4065	577	1589	922	45	1738	2978
Al	4561	1075	2390	933	70	2700	3130
Si	6920	859	2524	1685	169	2330	4863
Ca	2978	362	1378	1112	20	1550	2203
Fe	4200	1909	7042	1809	211	7874	3003
Co	4031	1928	7740	1768	209	8900	2905
Ni	4047	1834	7889	1726	207	8908	2796
In	2337	561	7015	430	11	7310	766
Cu	3440	1374	8089	1357	130	8920	2159
Zn	2850	817	6552	693	108	7140	2148
Ga	2873	724	5900	303	10	5910	749
Ge	2693	631	5487	1210	89,6	5323	2057
Ag	2790	955	9329	1234	83	10490	1658
Cd	2256	637	7997	594	51	8650	1446
Sn	2464	586	6973	505	50	7310	1400
Sb	1900	382	6077	904	55	6697	1540
Ba	1331	273	3343	1002	13	3510	1020
La	2030	728	5940	1193	37	6146	1443
Ce	1694	750	6550	1071	34	6689	1318
Pr	1926	716	6500	1204	37	6640	1380
Yb	1274	320	6720	1097	24	6570	1013
Ta	3303	2083	14353	3287	186	16650	2082
Pt	3053	1746	18909	2042	168	21090	1924
Au	2568	1162	17346	1336	78	19300	1536
Sc	4272	939	2680	1812	74	2985	3039
Ti	4309	1475	4141	1943	116	4507	3061
V	4255	1856	5340	2175	128	6110	2831
Y	3258	872	4180	1799	64	4472	2263
Zr	3648	1463	5650	2125	68	6511	2406

Metal	c (m/s)	γ_{LV} (mJ/m^2)	P_{lm} (Kg/m^3)	T_f (K)	E (GPa)	ρ_{sm} (Kg/m^3)	V_{RM} (m/s)
Nb	3385	1757	7830	2740	105	8570	2406
Pb	1821	471	10587	601	16	1146	2118
Pd	2657	1482	10495	1825	117	12023	742
Hf	2559	1517	11550	2500	78	13310	1789
Nd	2212	685	6890	1289	41	6800	1503
Sm	1670	431	7420	1345	50	7353	1411
Eu	1568	264	5130	1090	18	5244	1301
Gd	2041	664	7790	1585	55	7901	1083
Tb	2014	669	8050	1630	56	8219	1537
Dy	1941	648	8370	1682	61	8551	1525
Ho	1919	650	8580	1743	65	8795	1561
Er	1867	637	8860	1795	70	9066	1592
Lu	2176	940	9750	1936	69	9841	1426

Table 1.
Experimental sound velocities, c*, surface tensions,* σ_m*, densities,* ρ_{lm} *of different liquid metals at the melting temperature, elastic moduli,* E*, solid density,* ρ_{sm}*, and calculated Rayleigh velocities,* V_R *of these metals at solid state.*

3. Relationship between the properties of metals in solid and liquid states

Analytical study has been proposed to express the relation between experimental sound velocities of liquid metals at the melting temperature, *c*, and determinate

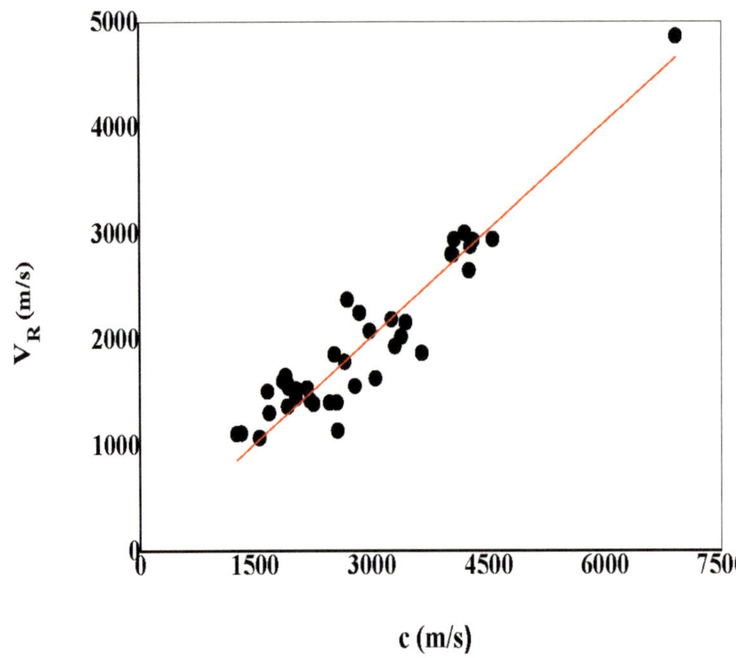

Figure 1.
Correlation between experimental sound velocities of liquid metals and Rayleigh velocities of these metals in solid state [20].

acoustic velocities, V_R, of these metals at solid state by SAM technique. The variation of V_R-values as function of c was made; it shows a linear increase of V_R with c increasing. Simple fitting was made and resulted in a well-defined linear correlation between the quantities, as can be seen in **Figure 1**.

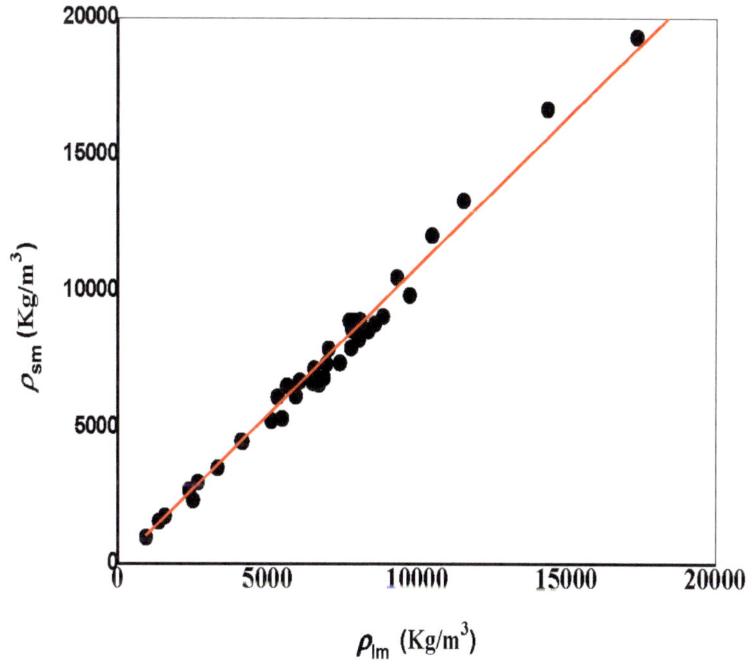

Figure 2.
Correlation between liquid and solid densities of metals [20].

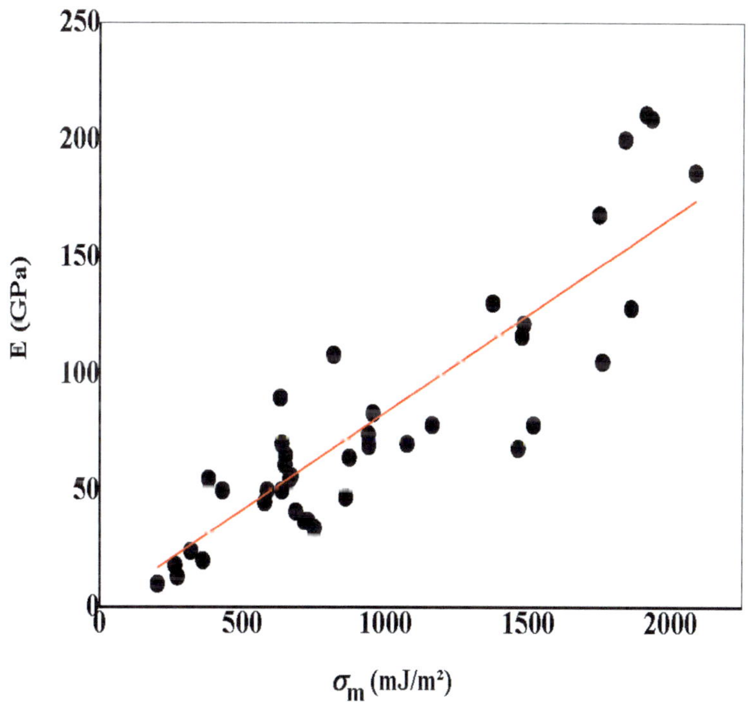

Figure 3.
Correlation between Young's moduli and surface tension of liquid metals [20].

Relationship between these parameters can be quantified by the following equation:

$$V_R = 0.674\, c \tag{5}$$

One can see also a clear tendency between the liquid metals densities, ρ_{lm}, with that of these metals at solid state, ρ_{sm}, as can be seen in **Figure 2**.

The relationship that expresses this tendency can take the following form:

$$\rho_{sm} = 1.088\, \rho_{lm} \tag{6}$$

A close comparative between one of very important properties of liquid metal, which is the surface tension, σ_m, and Young's moduli, E, values shows a linier dependence between these parameters, as can be seen in **Figure 3**.

To quantify the relationship between elastic moduli and surface tension, a simple plot was made; a linear correlation is defined, that it can be written as:

$$E = 0.083\, \sigma_m \tag{7}$$

The importance of the Eqs. (5)-(7) lies in the prediction of acoustic parameters from liquid to solid states of metals and vice versa.

4. Determination of adhesion energy in liquid metal/ceramic systems

Very recently, an electro-acoustical model [15] has been proposed to estimate and interpreted the work of adhesion of non-reactive liquid metal/ceramic substrates systems in terms of the Rayleigh velocity of acoustic wave propagation in surface of all types of corresponding ceramic substrates, V_{RC}.

In this model, several metals are considered (Au, Cu, Sn, Ga and Ag) on a great number of ceramic substrates (AlN, Al_2O_3, BN, CoO, Er_2O_3, Ho_2O_3, Lu_2O_3, MgO, NiO, SiC, SiO_2, TiC, TiO, TiO_2, Ti_2O_3, Y_2O_3, Yb_2O_3, ZnO and Zr_2O_3). The characteristics of all ceramic materials: energy gap, E_g [21] density, ρ_C, Young's modulus, E_C, and Rayleigh velocities [19] are listed in **Table 2**.

Ceramics Substrate	E_g (eV)	ρ_C (kg/m^3)	E_C (GPa)	V_{RC} (m/s)
AlN	5.6	3260	318	5616
Al_2O_3	7.1	3980	330	5650
BN	8.1	3487	34	1834
CoO	0.5	9423	281	2871
Er_2O_3	3.2	8651	179	2633
Ho_2O_3	3.9	8414	175	2639
Lu_2O_3	4.0	9423	204	2691
MgO	7.3	3580	310	5297
NiO	2.5	6670	420	6205
SiC	3.3	3210	393	6714
SiO_2	7.9	2600	75	3678
TiC	0.3	4940	400	5370

Ceramics Substrate	E_g (eV)	ρ_C (kg/m^3)	E_C (GPa)	V_{RC} (m/s)
TiO	0.0	4950	387	3960
TiO_2	3.1	4230	315	4917
Ti_2O_3	0.1	4468	118	4411
Y_2O_3	5.5	5030	176	3398
Yb_2O_3	1.4	9293	229	2677
ZnO	3.4	5606	125	2730
ZrO_2	8.0	5600	244	3781

Table 2.
Characteristics of investigated ceramic materials: energy gap, E_g*, density,* ρ_C *and Young's modulus,* E_C*, and determined Rayleigh velocities,* V_{RC}.

The variations of work of adhesion on Rayleigh velocity for different ceramic substrate, V_{RC}, in contacting with different non-reactive metals (Au, Cu, Sn, Ga and Ag) are investigated. In this study, some published data on wok of adhesion for different metals/ceramics systems are considered [12, 13, 22–34].

In the first time liquid gold/ceramic combinations are taken the obtained results are illustrated in **Figure 4**.

In order to generalize the above observations obtained with liquid Gold/ceramic systems and to put into evidence the results reproducibility, several other nonreactive metals deposed in different ceramic substrates are considered, i.e., (Cu, Sn, Ga and Ag):

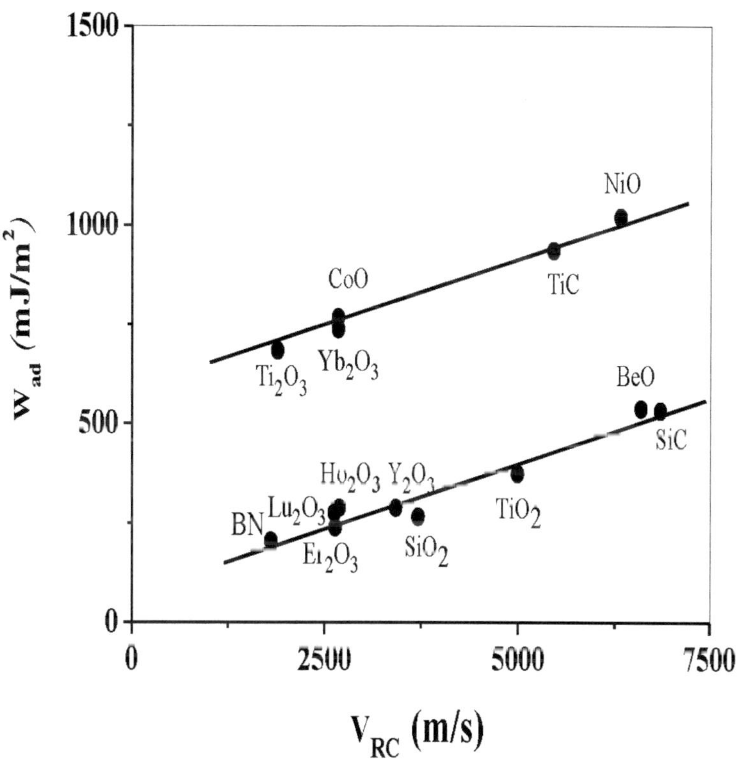

Figure 4.
Work of adhesion as function of calculated Rayleigh velocities of different ceramic substrates in contacting with gold [15].

The obtained results are illustrated in **Figure 5** in terms of work of adhesion as a function of ceramic Rayleigh velocities in contacting with several non-reactive metals. All the curves show the same behavior: the work of adhesion increases linearly with increasing V_{RC}. However, two sets of linear dependences are distinguished that are regrouped according to the band gap energy of the ceramic substrate, as discussed below.

The dependence of W_{ad} on V_{RC} (Au) is quantified via curve fitting, (lines in **Figures 4** and 5). We distinguish two parallel dependences for gold/ceramic substrate systems: for higher energy values (upper curve) the linear variation is found to be of the form:

$$W_{ad}(Au) = 0.07V_{RC} + 553 \quad (8)$$

Whereas, for small energy values (lower curve), the linear dependence is found to be of the form:

$$W_{ad}(Au) = 0.07V_{RC} + 76 \quad (9)$$

Moreover, it should be noted that the same behavior of two parallel lines is obtained for all metal/ceramic systems. Therefore, all curves have a same slop not only for small gap materials but also for large gap ceramics; the general expression takes the form:

$$W_{ad}\ (Me) = 0.07\ V_{RC} + C \quad (10)$$

$$W_{ad}\ (Me) = 0.07\ V_{RC} +' C \quad (11)$$

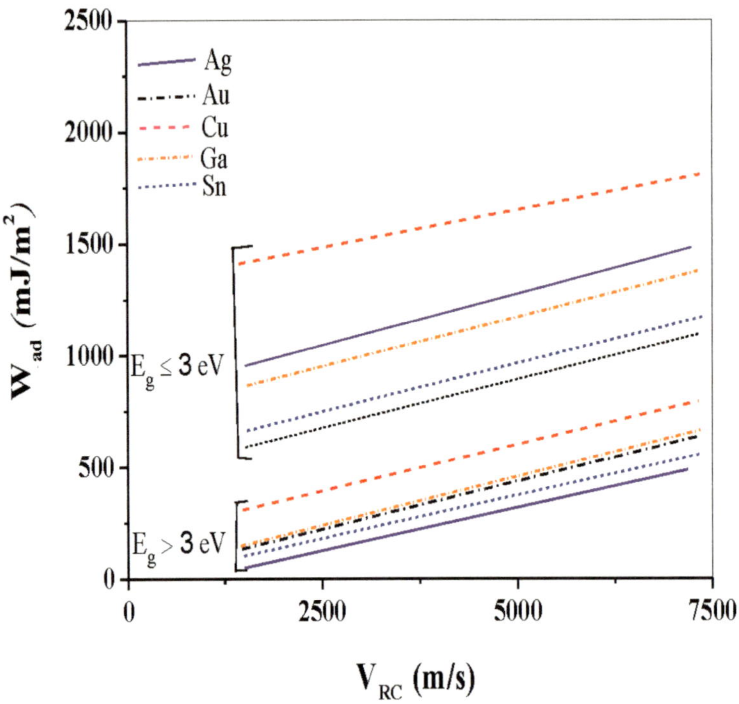

Figure 5.
Work of adhesion as function of calculated Rayleigh velocities of different ceramic substrates in contacting with several metals [15].

where the subscript, (Me), represents any given investigated nonreactive liquid metal (Ag, Au, Cu, Ga and Sn), C and Ć are characteristic constants for each metal/ceramic combination.

The exact corresponding values of characteristic constants C (for small gap ceramic materials) and Ć (for large gap ceramic materials) of several liquid metal/ceramic systems are giving in the **Table 3**.

The similar dependence (with the same slope equal to $0.07V_{RC}$) is indicative of the existence of the same mechanism responsible for this behavior. However the existence of two parallel dependences for every system is due to the energy band structure of the ceramic materials in particular the energy gap (**Table 1**). A close analysis of **Figure 5** and the E_g column clearly shows that the upper set of curves corresponds to solid ceramic materials with small energy gaps ($E_g \leq 3$ eV), whereas the lower ensemble of curves represents ceramic materials with large energy gaps ($E_g > 3$ eV).

In fact, solid materials with small band gaps behave as conductors ($E_g \rightarrow 0$) or semiconductors ($E_g \leq 3$ eV). In this case, it was reported [35] that the high adhesion energy values of same metal/ceramic systems are associated with high electron density of metals and low band gap energy of solids ceramics. The interfacial adhesion between a metal and a ceramic crystal is assured by the electron transfer [12], it is interesting to define an interfacial propriety represents the minimum energy needed for appearance of a limit number of interfacial bonds responsible for generating of the adhesion between the metal and the ceramic, this energy is caused by Van der Waals interaction, W_{VDW}. The intensity of the electron transfer at small band gap solid ceramic is increased because of its wealth by the free charges inside and the chemical equilibrium contribution $W_{chem\text{-}equil}$ taking place.

For large band gaps, there will be practically a small number of free charges inside in the ceramic crystal. As a result, the chemical equilibrium contribution Wchem-equil, to the adhesion energy is negligible. Consequently, the adhesion energy is approximately resulted by from the Van der Waals interaction [12].

The Van der Waals contribution of adhesion energy rested constant and proportional with Rayleigh velocity of ceramic materials whether it is the band gap energy, for the first time it is determined exactly as follows:

$$W_{VDW} = 0.07\,V_{RC} \tag{12}$$

The determinate W_{VDW} energy values for different metal/ceramic systems depend directly on the choice of various parameters appearing in Eq. (3). For example, Mc Donald and Eberhart [36] calculated W_{VDW} values equal to 500 ± 150 mJ/m^2 for different metal/alumina systems, that in our model and for the same system we have found W_{VDW} values equal to 396 mJ/m^2. While Naidich [13] found W_{VDW} values of 350 150 mJ/m^2 for metal/oxide ceramic systems, this

Metals	***C*** **(mJ/m^2)**	***Ć*** **(mJ/m^2)**
Ag	991	14
Au	533	76
Cu	1309	228
Ga	863	78
Sn	602	37

Table 3.
C and Ć values of different liquid metal/ceramic system.

confirms the compatibility between our proposed model and other model of W_{VDW} estimation.

For small gap ceramic materials, the characteristic constant C of Eq. (10) represents $W_{chem\text{-}equil}$ contribution, this energy is relatively important compared to W_{VDW} energy, it represents another interfacial property responsible for putting the stability and the perfection to the interface between metal and ceramic. The good convergence in $W_{chem\text{-}equil}$ values for a given metal/small gap ceramics could be explained by the fact that for (E_g < 3 eV), here will be a big density of inside in the ceramic crystal and consequently height electron transfer.

In this work, an analytical approach [20] is adopted to express the relation between experimental sound velocities of liquid metals, c, at the melting temperature and determinate Rayleigh velocity of these metals at solid state, V_{RM}, by SAM program. Hence, V_{RM} is expressed in terms of c, as we recently reported [20].

$$V_{RM} = 0.674\, c \tag{13}$$

The determinate chemical equilibrium energy, $W_{chem\text{-}equil}$ of metal/small band gap ceramic system by Eq. (10) are summarized in **Table 4**.

The variations of chemical equilibrium energy on normalized Rayleigh velocity, (V_{RM}/z) for different bulk metals in contacting with several small band gap ceramic materials are investigated, where z is number of coordination's of each metal atom. In this investigation, we consider Eq. (10) to determine $W_{chem\text{-}equil}$ and some published wok on adhesion energy for different metals/ceramics systems [12, 13, 22–34]. The obtained results are presented below.

The dependence of $W_{chem\text{-}equil}$ on (V_{RM}/z) is quantified via curve fitting, (line in **Figure 6**). We distinguish dependence for liquid metal/small band gap ceramic substrate systems: the linear variation is found to be of the form:

$$W_{chem-equil} = (1.3/z)V_{RM} \tag{14}$$

So, the chemical equilibrium contribution of adhesion energy in metal/ceramic system is related directly to the Rayleigh velocity of metals.

For large gap ceramic materials, the discrepancy in Ć values for a given metal/large gap ceramics could be explained by the fact that for (E_g > 3 eV), here will be a

Metals	$W_{chem\text{-}equil}$ (mJ/m^2)	Z
Ag	991	2
Al	1269	3
Au	553	3
Cu	1309	2
Co	1341	3
Fe	1276	3
In	723	3
Ni	1193	3
Ga	863	3
Sn	602	4

Table 4.
Determinate chemical equilibrium energy, $W_{chem\text{-}equil}$ *of metal/small band gap ceramic system and the number of coordination's of each atom of metal, z.*

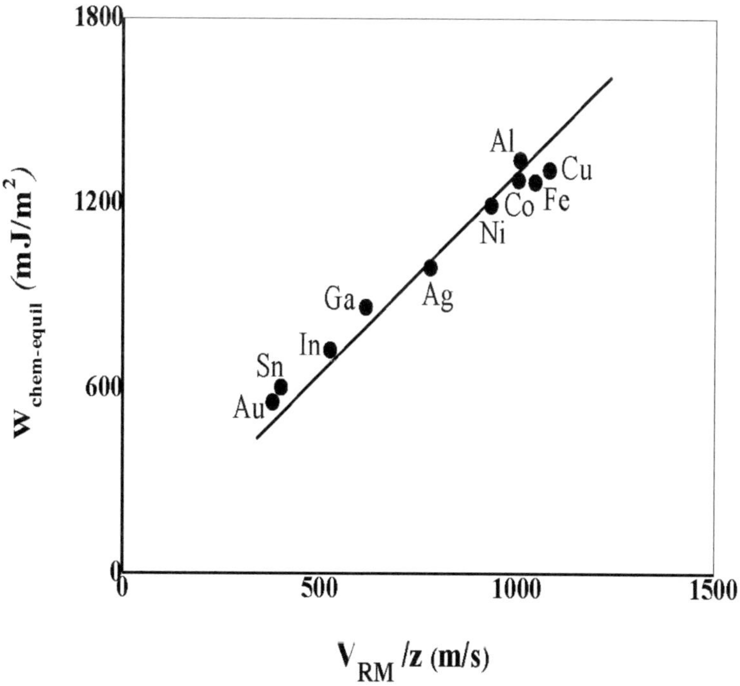

Figure 6.
Chemical equilibrium energy of metal/small band gap ceramic system as function of normalized Rayleigh velocities of different bulk metals [15].

smaller density of inside in the ceramic crystal (practically no free charges inside) and consequently the electron transfer at metal/ceramic interfaces cannot be important [2]. As a result, the characteristic constant Ć values are negligible compared to W_{VDW} energy and/or especially to $W_{chem\text{-}equil}$ energy.

Therefore, the general expression of adhesion energy takes the form:

a. For small gap ceramic materials:

$$W_{ad}\ (Me) = 0.07\ V_{RC} + (1.3/z)\ V_{RM} \tag{15}$$

b. For large gap ceramic materials:

$$W_{ad}\ (Me) = 0.07\ V_{RC} + W_{negl} \tag{16}$$

The importance of the deuced relation lies in its applicability to all investigated metal/ceramic systems. It could be extended, through familiar relations, to other acoustic parameters. Similar results for longitudinal and transverse velocities were obtained. Moreover, preliminary results for elastic constants (Young's modulus and shear modulus) are very satisfying.

5. Conclusions

In this work, an interfacial phenomenon between liquid metals and ceramic substrates has been investigated. Moreover, same liquid metal characteristics (sound velocity propagation in liquid metal, liquid density and surface tension) were predicted by the metal characteristics in solid state (Rayleigh velocity, solid density and Young's modulus). Adhesion energy terms in metals/ceramic systems were determined by using an electro-acoustic model. It was shown that the adhesion energy increases linearly with Rayleigh velocity of ceramic substrates for all types of ceramics. Van der Waals term of adhesion energy was deduced only depends on Rayleigh velocities of ceramic. On the other hand, the chemical equilibrium term was deduced strongly depends on the energy gap of the ceramics materials: it was higher for small band gap ceramic materials and depends on Rayleigh velocities of metals, for the opposite case it was deduced negligible. These universal relations that could be extended to other acoustic parameters are applicable to all metal/ceramic combinations.

Author details

Hadef Zakaria* and Kamli Kenza
Department of Physics, Faculty of Sciences, University 20 Août 1955-Skikda, Skikda, Algeria

*Address all correspondence to: zaki-hd2013@yahoo.fr; z.hadef@univ-skikda.dz

References

[1] Gordon I, Van Gestel D, Van Nieuwenhuysen K, Carnel L, Poortmans J. Thin-film polycrystalline silicon solar cells on ceramic substrates by aluminium-induced crystallization, Thin Solid Films, 2005;487: 113-117. https://doi.org/10.1016/j.tsf.2005.01.047

[2] Tabrizia AA, Pahlavan A. Efficiency, improvement of a silicon-based thin-film solar cell using plasmonic silver nanoparticles and an antireflective layer, Optics Communications, 2020; 454: 124437. https://doi.org/10.1016/j.optcom.2019.124437

[3] Yan D, Phang SP, Wan YM, Samundsett C, Macdonald D, Cuevas A. High efficiency n-type silicon solar cells with passivating contacts based on PECVD silicon films doped by phosphorus diffusion, Solar Energy Materials and Solar Cells, 2019;193: 80-84. https://doi.org/10.1016/j.solmat.2019.01.005

[4] Shao X, XiS, Li G, Liu Peng R, Li C, Chen G, ChenR. Longer hydrogenation duration for large area multi-crystalline silicon solar cells based on high-intensity infrared LEDs, Optics Communications, 2019; 450: 252-260. https://doi.org/10.1016/j.optcom.2019.06.011

[5] Legallais M, Thu T, Nguyen T, Cazimajou T, Mouis M, SalemB, Ternon C, Material engineering of percolating silicon nanowire networks for reliable and efficient electronic devices, Materials Chemistry and Physics, 2019;2 38: 121871, https://doi.org/10.1016/j.matchemphys.2019.121871

[6] Zhang X, Feng M, Zhao M, Zhang P, He C, Qi H, Han W, Guo F, Failure of silicon nitride ceramic flotation spheres at critical state of implosion, Applied Ocean Research, 2020;97: 102080 https://doi.org/10.1016/j.apor.2020.102080

[7] Bulyarskiy SV, The effect of electron-phonon interaction on the format,ion of reverse currents of p-n-junctions of silicon-based power semiconductor devices, Solid-State Electronics, 2019; 160: 107624. https://doi.org/10.1016/j.sse.2019.107624

[8] Wang J, Wu L, Chen X, Zhuo W, Avoiding blister defects in low-stress hydrogenated amorphous silicon films for MEMS sensors, Sensors and Actuators A: Physical, 2018; 276: 11-16. https://doi.org/10.1016/j.sna.2018.04.021

[9] Jantawong J, Atthi N, Leepattarapongpan C, Srisuwan A, Jeamsaksiri W, Sooriakumar K, Austin A, Niemcharoena S, Fabrication of MEMS-based capacitive silicon microphone structure with staircase contour cavity using multi-film thickness mask, Microelectronic Engineering, 2019; 206: 17-24. https://doi.org/10.1016/j.mee.2018.12.004

[10] Almuramady N, Borodich, Feodor MB, Goryacheva I, Torskaya EV, Damage of functionalized self-assembly monomolecular layers applied to silicon microgear MEMS, Tribology International, 2018; 129: 202-213. https://doi.org/10.1016/j.triboint.2018.07.049

[11] Frear D, Packaging Materials, Eds, Springer Handbook of Electronic and Photonic Materials, Springer Handbooks. Springer, Cham; 2017, https://doi.org/10.1007/978-3-319-48933-9_53

[12] Li JG, Chemical trends in the thermodynamic adhesion of metal/ceramic systems, Mater. Let.. 1995; 22: 169-174. https://doi.org/10.1016/0167-577X(94)00244-4

[13] Naidich YV, The wettability of solids by liquid metals, Progr. Surf.

Membr. Sci., 1981; 14:353-486. http://dx.doi.org/10.1016/B978-0-12-571814-1.50011-7

[14] Viktorov IA, Rayleigh and Lamb Waves, Plenum Press, New York, 1967.

[15] Kamli K, Hadef Z, Gacem A, Houaidji N. Prediction of Adhesion Energy Terms in Metal/Ceramic Systems by Using Acoustic Parameters, Metallophysics and Advanced Technologies, 2020; 42(5): 717–730. https://doi.org/10.15407/mfint.42.05.0717

[16] Blairs S, Correlation between surface tension, density, and sound velocity of liquid metals, Coll. Inter. Sci., 2006; 302: 312-314. https://doi:10.1016/j.jcis.2006.06.025

[17] Keene BJ, Review of data for the surface tension of pure metals, Int. Mat. Rev., 1993; 38: 157-192. https://doi.org/10.1179/imr.1993.38.4.157

[18] Crawley AF, Densities of Liquid Metals and Alloys, Int. Met. Rev., 1974; 19: 32-48. https://doi.org/10.1179/imtlr.1974.19.1.32

[19] Briggs GAD, Kolosov OV, Acoustic Microscopy, Oxford Univ. Press, 2nd Edition, New York, 2010. DOI:10.1093/acprof:oso/9780199232734.001.0001

[20] Hadef Z, Doghmane A, Kamli K, Hadjoub Z, Correlation between surface tension, work of adhesion in liquid metals/ceramic systems and acoustic parameters, Progress in Physics of Metals, 2018; 19(2): 198-223. https://doi.org/10.15407/ufm.19.01.198

[21] Strehlow WH, Cook EL, Compilation of Energy Band Gaps in Elemental and Binary Compound Semiconductors and Insulators, J. Phys. Chem. Ref. Data, 1973; 2: 163-199. https://doi.org/10.1063/1.3253115

[22] Eustathopoulos N, Sobczak N, Passerone A. Nogi K. Measurement of contact angle and work of adhesion at high temperature, Mater. Sci., 2005; 40: 2271-2280. https://doi.org/10.1007/s10853-005-1945-4

[23] Naidich YV, Zhuravlev VS, Frumina NI, Wetting of rare-earth element oxides by metallic melts, Mater. Sci., 1990; 25: 1895-1901. https://doi.org/10.1007/BF01045739

[24] Li JG, Wettability of Solid Inorganic Materials by Gold, Scripta Metall. Mater., 1994; 30: 337-342. https://doi.org/10.1016/0956-716X(94)90385-9

[25] Taranets NY, Naidich YV, Wettability of aluminum nitride by molten metals, Powder Metall. Met. Ceramics, 1996; 35: 74-78. https://doi.org/10.1007/BF01328834

[26] Liu GW, Muolo ML, Valenza F, Passerone A, Survey on wetting of SiC by molten metals, Ceram. Inter., 2010; 36: 1177-1188. https://doi.org/10.1016/j.ceramint.2010.01.001

[27] Kida M, Bahraini M, Molina JM, Weber L, Mortensen A, High-temperature wettability of aluminum nitride during liquid metal infiltration, Mater. Sci. Eng. A, 2008; 495: 197-202. https://doi:10.1016/j.msea.2007.12.050

[28] Naidich YV, About liquid metal/ceramic interface interaction mechanism and mode of a new intermediate compound formation, Curr. Opi. Sol. Sta. Mater. Sci., 2005: 161-166. https://doi.org/10.1016/j.cossms.2005.11.001

[29] Li JG, Wetting and Interfacial Bonding of Metals with Ionocovalent Oxides, J. Amer. Ceram. Soc., 1992; 75: 3118-3126. https://doi.org/10.1111/j.1151-2916.1992.tb04396.x

[30] Li JG, Hausner H, Contact angle and work of adhesion isotherms of silicon-tin alloys on monocrystalline silicon carbide, Mater. Let,, 1991; 11: 355-357.

https://doi.org/10.1016/0167-577X(91)90133-Q

[31] Li JG, Wetting and interfacial bonding in liquid metal/solid ceramic systems, Comp. Interf., 1993; 1: 37-53. https://doi.org/10.1163/156855493X00301

[32] Li JG, Hausner H, Wetting and adhesion in liquid silicon/ceramic systems,Mater. Letters, 1992; 14: 329-332. https://doi.org/10.1016/0167-577X(92)90047-N

[33] Chatain D, Rivollet I, Eustathopoulos N, Adhésion thermodynamique dans les systèmes non-réactifs métal liquide-alumine, Chim. Phys., 1986; 83: 561-567. https://doi.org/10.1051/jcp/1986830561

[34] Sotiropoulou D, Nikolopoulos P, Work of adhesion in ZrO_2-liquid metal systems, J. Mater. Sci., 1993; 28: 356-360. https://doi.org/10.1007/BF00357807

[35] Li JG, Role of electron density of liquid metals and band gap energy of solid ceramics on the work of adhesion and wettability of metal-ceramic systems, Mater. Sci. Let., 1992; 11: 903-905. https://doi.org/10.1007/BF00729089

[36] Mc Donald JB, Eberhart JG. Adhesion in aluminum oxide-metal systems, Trans. AIME, 1965; 233: 512-517.

Section 3

Surface Modification

Chapter 5

Microstructural and High Temperature Wear Characteristics of Plasma Transferred Arc Hardfaced Ni–Cr–Si–B-C Alloy Deposits

S. Gnanasekaran, Samson Jerold Samuel Chelladurai, G. Padmanaban and S. Sivananthan

Abstract

Due to the tough working environments, wear damage to nuclear reactor components is frequent. Usually, nuclear elements run at 573 k to 873 k. The feed water controller valves, used for the thundering of coolant flow, wear out faster among the reactor components. Austenitic stainless steels, using different methods for hardfacing, improve wear resistance to the cobalt and nickel alloys. Nickel based hardfacing is more resistant to wear than cobalt based hardfacing at high temperatures thanks to the solid oxide layers. Austenitic stainless-steel substrates generally favor nickel-based hardfaced (Ni–Cr-Si–B-C) over cobalt-driven hardfacing because this reduces radiation-induced nuclear activity. A well-known surface method for depositing nickel hardfacing, minimal dilution, alloys is the Plasma Transfer Arc (PTAs) weld technique. In this study the Ni-based alloy is hardfaced over a 316 L (N) ASS substratum with PTA hardfacing, for a dense of approximately 4–4.5 mm. The substrates and deposits were tested at different temperatures with a pin on disc wear (room temperature, 150 and 250°C).When grinding with 1000 grain SiC abrasive paper, the wear test samples were polished to the roughness value (Ra) of less than 0,25 m.The deposit showed a variety of wear mechanisms regarding the test temperature. Using friction and wear values and wear analysis, the wear mechanisms were determined. There was a considerable wear loss at room temperature (RT).At 423 K operating heat, mild ploughing at short sliding distances and tribo-oxidation were carried out with increasing sliding time.The primary wear mechanism was adherence at the time of operating temperature at 623 K, but as the sliding distance widened, tribo-oxidation improved. In combination with a working hardened substrate, the formation of an oxide layer could significantly reduce the wear loss of nickel-based alloys.

Keywords: Austenitic stainless steel, PTA hardfacing, Wear, Microstructure

1. Introduction

Austenitic type 316 L (N) is commonly used in fast-breeder (FBR) reactors at temperatures between 573 K and 874 K as structural material. Austenitic stainless steels are very resistant to wear under dry sliding conditions [1]. Nickel and cobalt alloys are recommended to be used as layer materials for deposition on austenitic stainless steels using different layering methods to increase wear resistance [2–5]. The soldering technique Plasma Transferred Arc (PTA) is known for depositing nickel and cobalt hard façade alloys with limited dilution on rust-resistant stainless-steel substrates. Nickel-based hard coating on austenitic stainless-steel substrates is commonly a suitable alternative to cobalt-based coating for use in the nuclear industry to minimize radiation-inducing activity [6–8].

Many laboratory experiments have been performed on nickel-based hardfacing alloys as these alloys are both corrosion and oxidation resistant. Kashani et coll. [9] contrasted the use efficiency of room temperature(RT) and high temperature (HT) nickel and cobalt-based hard-facing coatings (823 K). Due to the lightweight oxide layers, they were found to be more wear-resistant to nickel coatings than to cobalt coatings at high temperatures [10]. The wearing conduct of RT, 373, 473, 498, 523 and 623 K of a Deloro 50 hardfacing alloy was investigated. They found the serious Space temperature wear loss (RT). However, the wear loss decreased to near zero at 623 K as the test temperature rose. They suggested to use a nickel-based hardfacing alloy for applications in nuclear power plants where the working temperature is high enough to produce oxidative wear. Berns et al. [11, 12] Hardness test during slipping at high test temperatures witnessed coating operation. In combination with a working hardened substrate, the forming of an oxide layer could greatly minimize the wear loss of nickel-based alloys. Coating sliding can show a variety of wear appliances at different trial temperatures and ranges [13]. Consequently, room and high temperature wear behaviors, using the high-temperature pin-on-disk wear test process, of a Colmonoy 5 hardfacing alloy dropped on a 316 L (N) stainless steel substratum were examined.

2. Experimental work

2.1 Substrate (Base metal) and hardfaced powder (colmonoy-5) properties

AISI 316 LN is an austenitic nuclear-grade stainless steel widely used in valves, valve cones, spins and other structural grain stalks. A vacuum spectrometer determined the chemical composition of the base metal. Sparks of the base metal were ignited at various locations. The resulting spectrum has been studied to estimate alloying components. **Tables 1** and **2** illustrate the chemical compositions of base metal and powder. The content of the foundation was rolling 12 mm thick. The experiments were carried out using a semi-automated PTA tool. The tests were performed by creating a negative electrode charged layer (DCEN). As blinding gas and as powder feed gas, pure argon (99.9 percent) gas was used.

Table 3 displays the configured parameters of the PTA hardfacing mechanism for this study. The deposits were faced with four degrees of principal arc current, while all other parameters were kept stable. The hardfaced samples are seen in **Figure 1**. The deposit was reduced for the study of metallography into small samples after hard-facing. The hardness of the cross section of the repository was tested with a 0.05 sec, with a dwell time of 15 s and a Vickers microhardness measuring press, (Make: SHIMADZU, Japan; Model: HMV2T).

C	Ni	Cr	Mo	Si	Mn	Cu	Nb	S	P	W	Fe
0.020	12.55	17.27	2.35	0.29	1.69	0.047	0.02	0.027	0.026	0.03	**Bal**

Table 1.
Chemical composition wt% of 316LN (Substrate).

C	Fe	Cr	Si	B	O	Ni
0.41	3.10	10.44	4.02	2.26	0.03	**Bal**

Table 2.
Chemical composition of colmonoy-5 (Hardfaced powder).

S.No	Main parameters	Values
1	Main arc current (Amps)	130
2	Traverse speed (mm/min)	160
3	Powder feed rate (grams/min)	35
4	Torch oscillation width (mm)	10
5	Preheating temperature (°C)	400

Table 3.
PTA process parameters used for hardfacing.

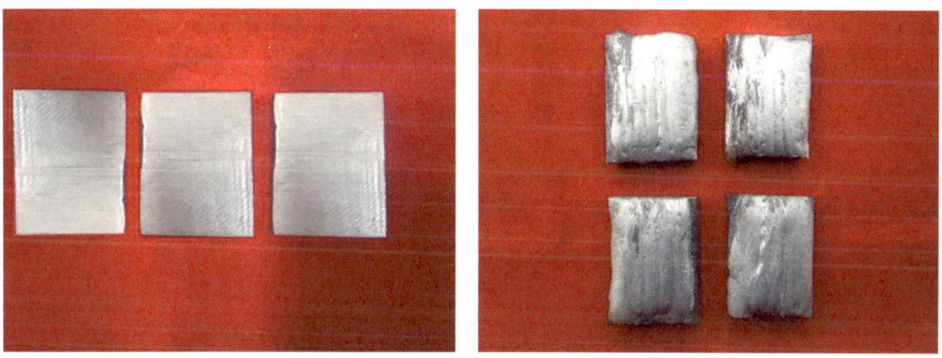

Figure 1.
Photograph of hardfaced specimens. (a) Before hardfacing. (b) After hardfacing.

Parameters	Values
Pin	D = 10 mm, L = 20 mm
Disc	D = 160 mm, W = 8 mm
Load	50 N
Velocity Range	1 m/s
Sliding distance	450 m, 800 m, 1200 m, 1600 m

Table 4.
Wear test parameters.

The specimens were separated to the required size and polished with different grades of emery paper for metallographic inspection. A typical reagent of 0.25 g, 20 ml ethanol and 1.25 ml of HCL were used to detect the microstructure of the specimens. The microstructural research has been carried out using the optical light-emitting microscope (OM) (Make: MEIJI, Japan; Model: MIL7100) (Metal Vision). Pin-on disc wear experiments were carried out at three different temperatures for the layer and deposit (room temperature, 423 K, and 623 K) (**Table 4**).

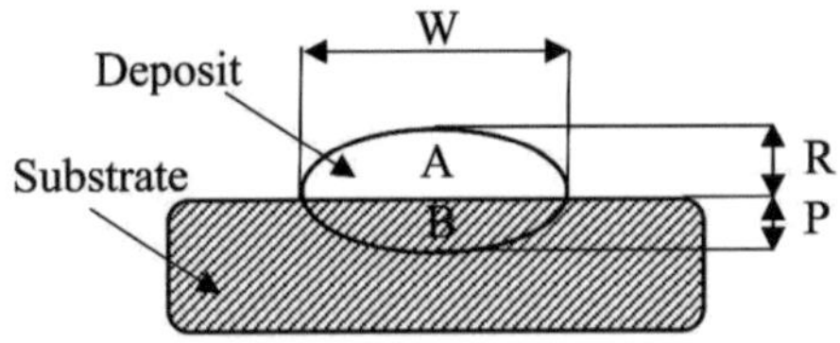

Where,

P - Penetration
R - Reinforcement
W - Width
A - Reinforcement area
B - Penetration Area

Figure 2.
Bead profile and bead geometries.

2.2 Dilution

The welding method with the least dilution is generally favored for tough applications. Dilution is the proportion of the base metal in a metal deposit. As the dilution increases, the amount of base metal in the welded metal deposit increases and vice versa. Hardfacing is achieved mainly to enhance the base metal (substrate) surface properties. Hardfaced surfaces usually outperform the wear, degradation and oxidation resistance of base metals. The surface properties do not change to the expected degree in higher dilution rates because of the inclusion of a larger volume of base metal. This section performed an experimental study to assess the dilution percentage at different power levels. After hardfacing, the specimens were selected from the centre of the deposit. Metallographic technique and specimens of 2% Nital were used to polish the surface of the material. In high-resolution scanners the bead profile and geometry as seen in **Figure 2** was visualized and registered. Dilution was calculated using the following expression.

$$\text{Dilution}\,(\%) = \frac{B}{A+B} \text{ x } 100\% \quad (1)$$

3. Results and discussion

3.1 Characterization of the deposits

At higher transferred arc currents, heat generation is higher and the material is melted after the powder excess heat. In addition, increased arc forces improve heat generation, which increases the penetration depth of the substrate material. The heat generation is lowers at lower transferred arc currents and the rest of the heat generation is used to melt the powder, allowing less heat to melt the base material after melting the powder. In addition, the arc force in this state is decreased and the penetration depth and superficial penetration decrease. Dendritic nickel growth nearly perpendicular to the deposit interface with a greater magnification has been detected (**Figure 3a**). The microstructure of the deposits includes -nickel dendrites with precipitation of interdendritic carbide. The deposit near the interface is significantly different from the top of the cover (**Figure 3b**).

Figure 4 displays the hard-faced deposit scanning electron micrograph consisting of the nickel solid solution phase dendrites and the eutetic mixture as interdendritic components. The average hardness of the deposit is approximately 530 HV, and is held at 1 mm from the interface. Basic metals have an average hardness of around 250 HV. The Colmonoy 5 coatings have averaged two times higher hardness relative to the substrate. The existence of an evenly dispersed blend of complex carbohydrates and borides causes the improved stiffness of the overlays (**Figure 5**).

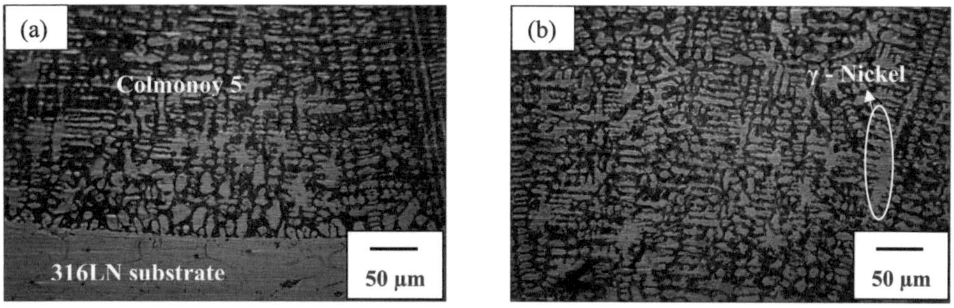

Figure 3.
Optical micrographs of (a) Interface with etching and (b) hardfaced deposit.

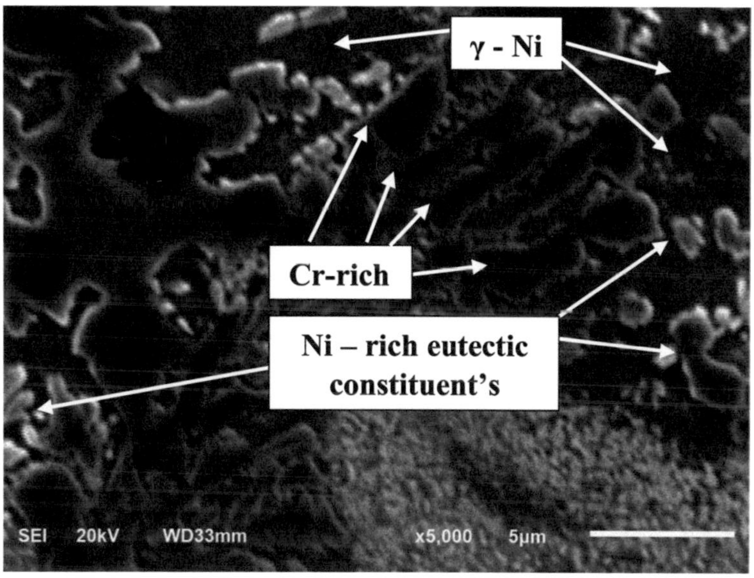

Figure 4.
Scanning electron micrograph - Cr-rich precipitates.

3.2 Dry sliding wear behavior

The apparel of the depot is seen in **Figure 6** due to test temperatures in different sliding lengths. Wear loss is evident as sliding distance increases and the test temperature decreases. RT inspections indicate further wear loss on the deposit. Wear loss decreased at 423 K and 60–75% at 623 K at higher test temperatures by 20–30 percent in comparison with RT samples. The estimated deposit wear ratios at RT, 423 K and 623 K are 3.01, 2.1 and 1.07 g/Nm. Due to the development of smoothly crushed oxide layers on sliding sides, apparel loss of the nickel-based hardfaced alloy may be reduced as stated in detail [14]. In the incidence of the oxygen atmosphere, oxide layers are more readily formed in contact areas that have been locally adjusted due to the superimposed operating temperature and frictional temperature effects. The highest wear loss (60–70%) for 1600 m of sliding reserve was found throughout the running age during RT experiments. The plan of harsh stringencies and the combination of apparel debris, which led to scratch and investing of all sliding sides, caused the severe wear during the run-in period [15].

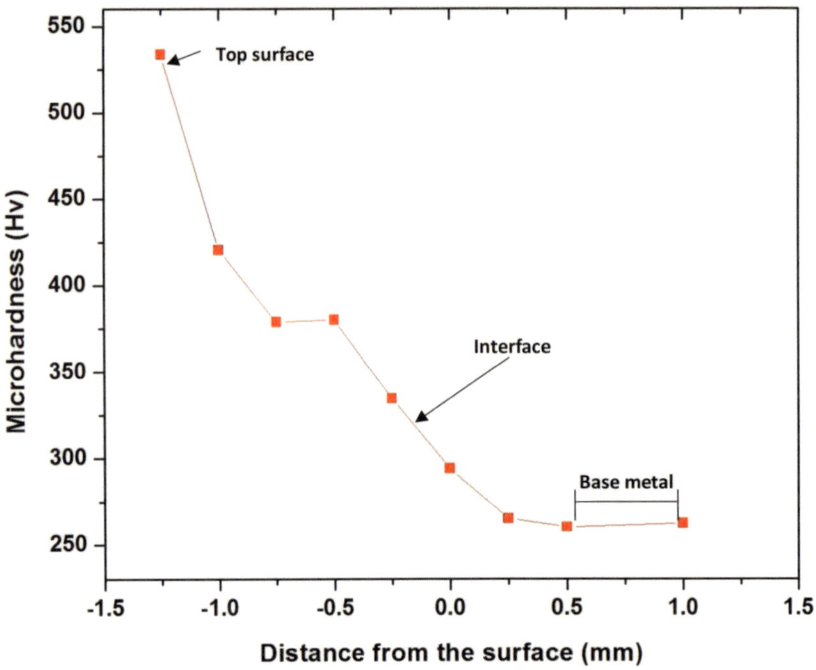

Figure 5.
Stability distribution between substrate and deposit.

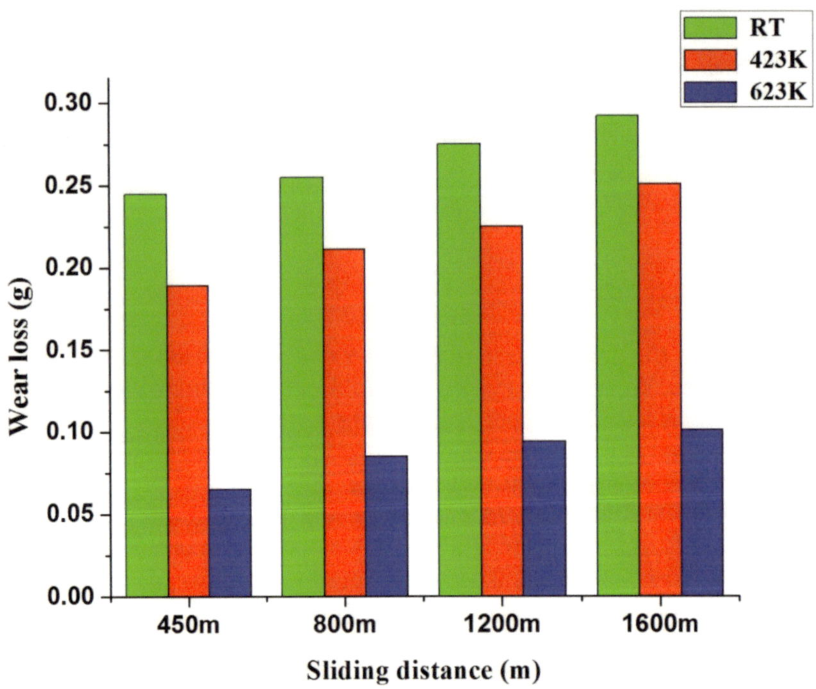

Figure 6.
Outcome of test temperature on wear loss of the deposit at unlike sliding distances.

3.3 Dry sliding friction behavior

The sliding distance is compared to the operational and stable friction coefficient (COF) morals during RT and at extraordinary temperatures in **Figure 7**. The COF development with a sliding distance is highly temperature dependent. At 623 K and 423 K COF is significantly less than RT. The oxidation level is inadequate at 423 K at higher temperatures and is also higher than the COF tests at 623 K. The COF bends showed different sections for running and stable periods throughout 350–450 meters of sliding distance. During the run-in period, COF at RT rose to 0.7, then dropped to 0.45. During the steady state condition, there is also a fluctuation between 0.45 and 0.55. At both operating and static cycles, COF decreased, with test temperatures increasing to 423 K and 623 K. During the operating cycle, however, COF at all test temperatures is greater than static COF. The average constant state COF at RT (0.50–0.55) reduced to 0.35–0.45 and 0.25–0.35 respectively with arise in the test temperatures to 423 K and 623 K. COF, especially after 400 m, is very unstable at 623 K. In order to temporarily shield surfaces from further impact damages, Stott et al. [16] characterize the part of tribo-layers on sleeveling sides. These protective layers of glaze are continually refilled and worn down as the previous layer slides. As seen in **Figure 7**, the growth of formation, wear and reform would lead to short-term COF changes of 623 K.

3.4 Wear mechanisms for 1600 m sliding distance

The area found for sliding on the RT is smooth, with only slight abrasion, and a delamination segment (**Figure 8(a,b)**). As can be seen in **Figure 8**, highly deformed

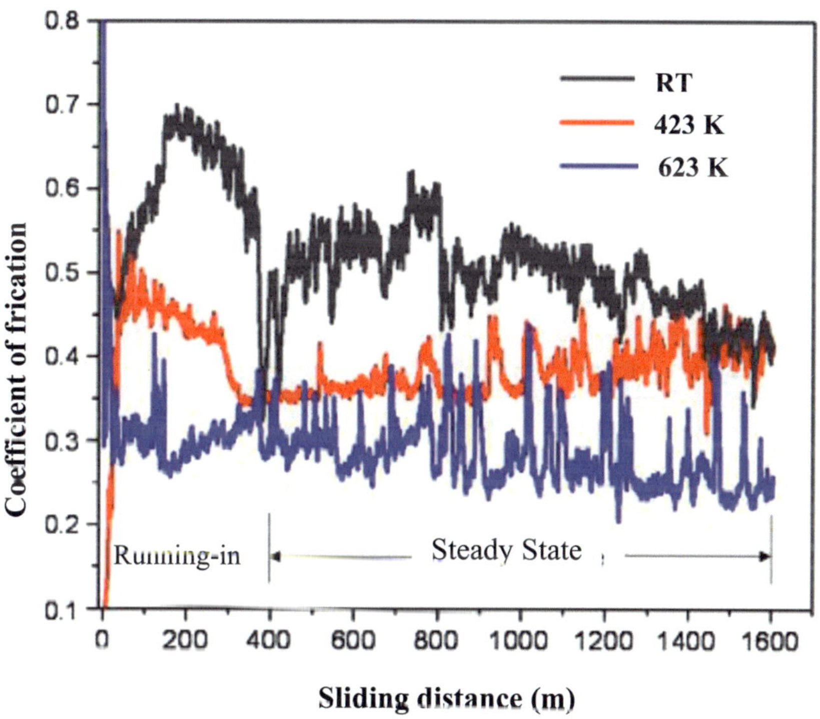

Figure 7.
Quantity of friction of the model tested at (a) RT (b) 423 K and (c) 623 K (sliding velocity 1 m/s and load 50 N).

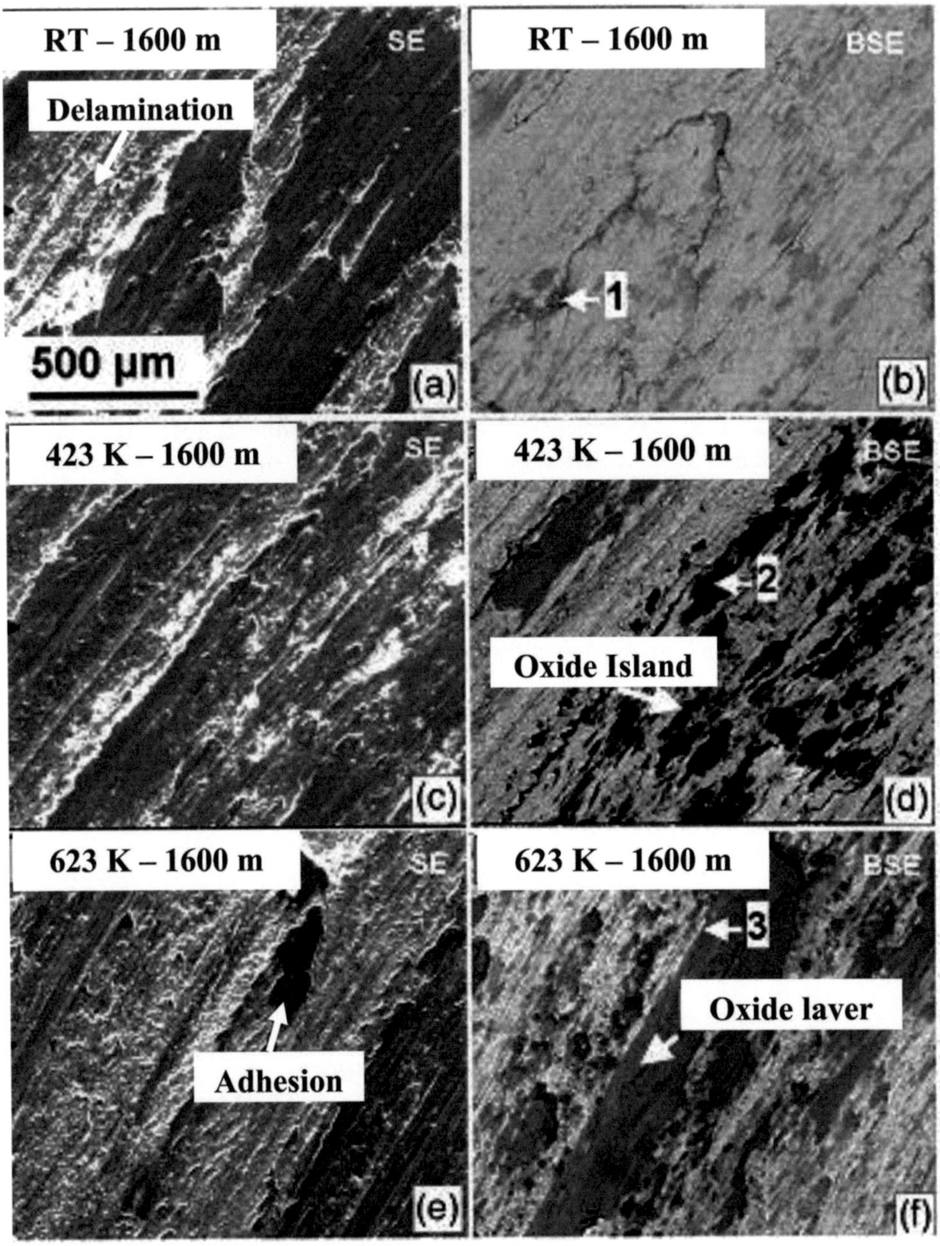

Figure 8.
SEM micrograph features of the long sliding distance worn surfaces (1600 m): (a–b) the smooth surface of the sliding RT shows delaminated portions with mild abrasive marks. (c-d) The area with mild abrasive marks of oxide islands at 423 K. (e–f) Severe 623 K coating of compacted oxide.

bits and surfaces with oxidation and minor scratches are present at this point in line with motions of the trailer (c,d). This is because the coating's spontaneous oxidation increases the output of oxide waste. The oxides on the worn area rise as the test temperature goes up to 623 K (**Figure 8(e,f)**). The wear path is oxidized more than the surface outside the contact area and therefore tribo-oxidation is required. The oxide layers are difficult enough to withstand wear and load [9].

Oxygen contained the worn surfaces, with the increase in temperature the oxygen content increased (denoted as 1, 2 and 3 in **Figure 8**). A discontinuous oxidized film covers the worn foundation. The oxide layer composition is identical to the bulk depot composition, but at higher test temperatures, the oxygen level remains high. The distribution of the oxide layer on the tracks is not uniform. Since the processes contained in diagram as shown in **Figures 9** and **10** were interesting in 623 K, instead of RT, an attempt was made to clarify. Surfaces are divided by a slight distance in contact with each other. As sliding begins and asperities become broken and more debris development occur (**Figure 9(b)**), the shaving load reaches its limit. The processing of waste helps to minimize the interaction with asperities and quickly produces considerable waste and closes the soil.

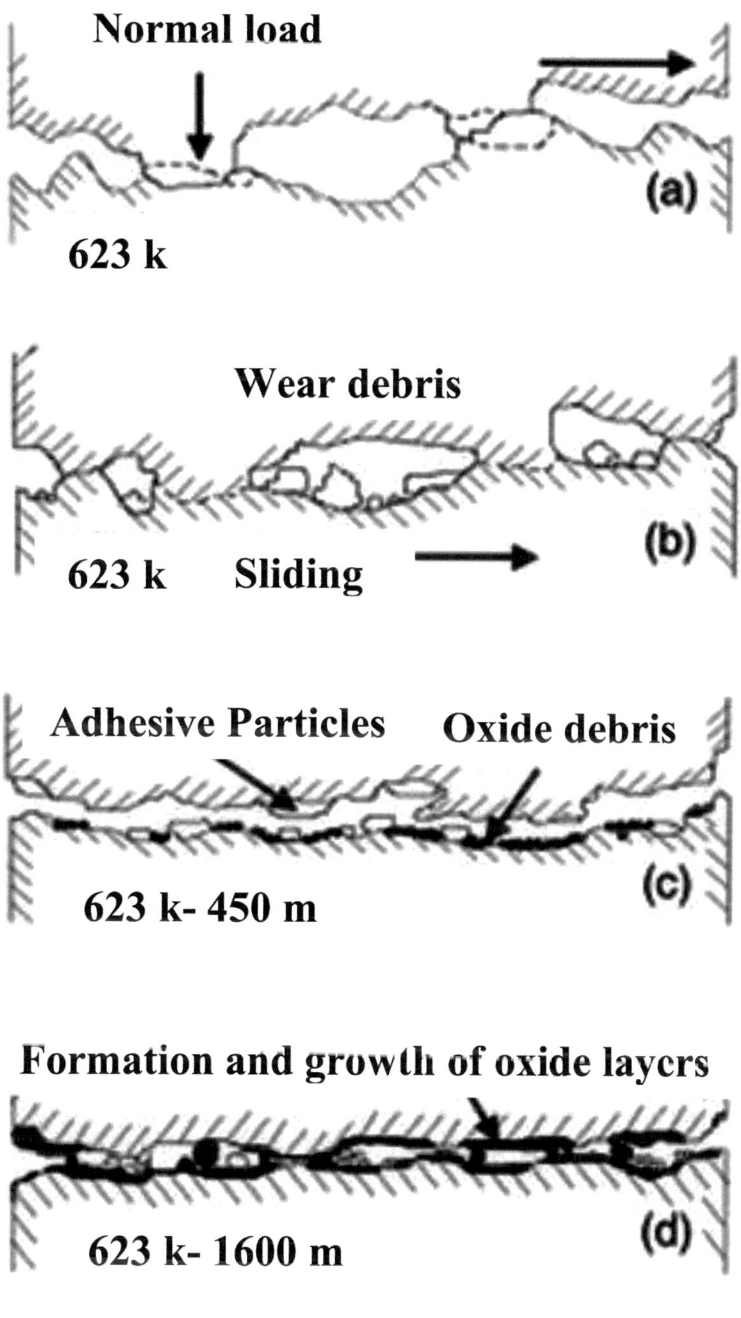

Figure 9.
Schematic representation of wear mechanisms of Colmonoy 5 deposit at RT [Ref - D. Kesavan et al. 2010].

The matrix will pluck in hard asperities, resulting in deep scratches (**Figure 9(c)**). In the other hand, debris is a third body particle that is abrasive, and causes abrasion on both directions. The slipping surfaces get smoother with time with the flat contact of the coating surface decreasing the more wear of the deposits (**Figure 9(d)**). **Figure 10(a–d)** shows the extensive wear-pathway processes at 623 K. The oxide layer prevents gradual surface wear at elevated temperatures due to its higher inherent hardening, preventing ploughing. In forming the oxide layer, the following processes are involved: Asperity fracture debris was concerned. The combination scrapping and oxidation of fine oxidized debris causes a fine tribo-layer to develop both on the

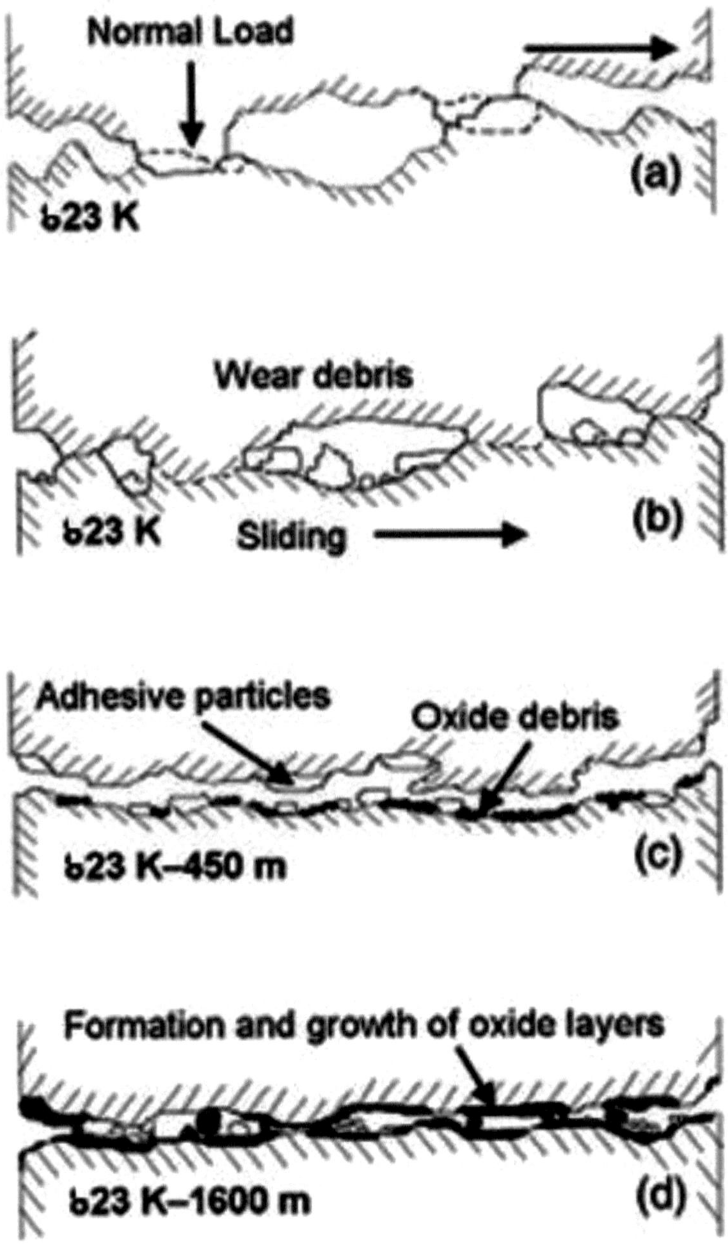

Figure 10.
Schematic representation of wear mechanisms of Colmonoy 5 deposit at 623 K [Ref - D. Kesavan eat al 2010].

surfaces of the pin and on the disc (**Figure 10(d)**). Further sliding is expected on the coating surface between tribo-layers before this tribo-layer forms. This prevents open contact between metal and metal.

4. Conclusions

1. AISI 316 L(N) austenitic stainless-steel substratum with PTA hardfaced technique was successfully deposited with nickel-based Colmonoy 5 powder at a thickness of four mm. The hardness of the deposit improved by an average of two relative to the substratum. The fact that the complicated carbides blend is spread uniformly causes the deposit to become more robust.

2. The wear loss and coefficient of friction (COF) of the deposit decrease with the rise of test temperature. The measuring temperature has a higher effect on the wear defeat of the coating than the slip space.

3. Understanding of wear loss It can be concluded that wear processes operating on the RT over the initial sliding distance is intense abrasion and tilting, according to the characteristics of the worn surface regions. In the other side, long sliding distances led to delamination and minor abrasion during wear.

4. At 423 K operating heat, mild ploughing at short sliding distances and tribo-oxidation were carried out with increasing sliding time.

5. The primary wear mechanism was adherence at the time of operating temperature at 623 K, but as the sliding distance widened, tribo-oxidation improved.

Author details

S. Gnanasekaran[1], Samson Jerold Samuel Chelladurai[2*], G. Padmanaban[3] and S. Sivananthan[4]

1 Mechanical Engineering, Sri Sakthi Institute of Engineering and Technology, Coimbatore, Tamil Nadu, India

2 Department of Mechanical Engineering, Sri Krishna College of Engineering and Technology, Coimbatore, Tamil Nadu, India

3 Centre for Materials Joining and Research (CEMAJOR), Department of Manufacturing Engineering, Annamalai University, Annamalainagar, Tamil Nadu, India

4 Mechanical Engineering, K. Ramakrishnan College of Engineering, Trichy, Tamil Nadu, India

*Address all correspondence to: samsonjeroldsamuel@skcet.ac.in

References

[1] K.L. Hsu, T.M. Ahn, D.A. Rigney, Wear 60 (1980) 13.

[2] R.L. Deuis, J.M. Yellup, C. Subramanian, Compos. Sci. Technol. 58 (1998) 299.

[3] O. Knotek, E. Lugscheider, H. Reimann, Thin Solid Films 64 (1979) 365.

[4] C.T. Kwok, F.T. Cheng, H.C. Man, Surf. Coat. Technol. 145 (2001) 206.

[5] Dawei Zhang, Xinping Zhang, Surf. Coat. Technol. 190 (2005) 212.

[6] M. Corchia, P. Delogu, F. Nenci, A. Belmondo, S. Corcoruto, W. Stabielli, Wear 119 (1987) 137.

[7] D.H.E. Persson, S. Jacobson, S. Hogmark, Wear 255 (2003) 498.

[8] A.K. Bhaduri, R. Indira, S.K. Albert, B.P.S. Rao, S.C. Jain, S. Asokkumar, J. Nucl. Mater.334 (2004) 109.

[9] H. Kashani, A. Amadeh, H.M. Ghasemi, Wear 262 (2007) 800.

[10] Seon Jin-Kim, Jun Ki-Kim, J. Nucl. Mater. 288 (2001) 163.

[11] A. Pauschitz, Manish Roy, F. Franek, Tribol. Int. 41 (2008) 584.

[12] H. Berns, Wear 181 183 (1995) 271

[13] K. Gurumoorthy, M. Kamaraj, K. Prasad Rao, A. Sambasiva Rao, S. Venugopal, Mater. Sci. Eng. A 456(2007) 11.

[14] G.W. Stachowiak, A.W. Batchelor, Tribology Series, vol. 24, Engineering Tribology

[15] Elsevier Science Publisher, Netherlands, 1999, p. 546.

[16] F.H. Stott, Tribol. Int. 35 (2002) 489.

Chapter 6

Fabrication of Multi-Layered Zn-Fe Alloy Coatings for Better Corrosion Performance

Ramesh S. Bhat

Abstract

Zn-Fe compositionally modulated multilayer alloy (CMMA) coatings were developed onto low carbon steel from acid sulphate bath; and their corrosion resistance was calculated using Tafel polarization and impedance methods. The deposit layers were formed galvanostatically by single bath technique (SBT), using square current pulses. An optimal configuration for the growth of most corrosion resistant Zn-Fe coating was proposed and discussed. At maximum switching cathode current density (SCCD) (2.0–5.0 A dm^{-2}), the deposit with 300 layers showed ~43 times superior corrosion resistance than the same thickness of monolayer coating. The improved corrosion resistance of multilayered coatings is due to small change in iron content, leading to change the phase structure of the alternate-layers of the alloy coatings. The surface morphology and structure of film and roughness of the deposit were assessed using Scanning Electron Microscopy and Atomic Force Microscopy. Thus, superior corrosion resistance of Zn-Fe multilayer coatings was used for industrial applications including defense, machinery and automobile etc.

Keywords: corrosion, electroplating, single bath technique, thickness, additives

1. Introduction

Electroplating, also known as electrodeposition, is the process of creating solid films from dissolved substances by changing their oxidation states with electricity. The electroplating technique is a flexible tool for improving the characteristics of a metal, alloy, or composite on the surface of a substrate. It has a wide range of uses in electronics, protective coatings, and a variety of other sectors of surface engineering. Many books are available that cover a wide range of issues, including surface preparation prior to deposition, atomistic reactions, electrodeposition thermodynamics and kinetics, crystal growth mechanisms, bath chemistry, and so on [1, 2]. Electrochemical polarization techniques and other instrumental approaches were used to investigate the electrodeposition of Zn-Fe alloys from a chloride-based electrolyte [3–6]. Using cyclic voltammeter (CV) and the steady state polarization approach, the behavior of Zn, Fe, and Zn-Fe alloy onto copper from acid chloride solutions containing EDTA and boric acid was examined [7]. The Zn^{+2} ions in the electrolyte prevent Fe from forming, while Fe^{+2} ions stimulate Zn formation. The codeposition of Zn and Fe was discovered to behave peculiarly [8–10]. Yang et al. [11] created a sulphate bath for the preparation of Zn-Fe alloy coatings. As a buffer and a complexing agent, citric acid and sodium citrate were utilized [12]. The interfacial pH was discovered to control the kinetics of metal deposition.

Materials with ultra-fine microstructures hold promise as new generation materials, and one such group of materials is compositionally modulated multilayer alloy (CMMA) coatings. It is made up of metal or alloy layers that alternate in thickness (a few nanometers). Multilayered deposits may have extraordinary and, unique functional qualities that are not possible to obtain in conventional metallurgical alloys. The corrosion resistance of CMMA coatings will be improved [13–17], such as giant magneto-resistance, increased hardness, wear and corrosion resistance, optical x-ray characteristics etc. The thickness of each layer in a CMMA coating, which should be as thin as feasible, is the most essential condition for displaying these qualities. Periodic modifications in the deposition circumstances can be used to provide a periodic change in the composition/structure of every layer [18]. Dry or wet techniques are used to create multilayered alloy coatings. There are several methods used for the production of CMMA coatings in dry process, including physical vapor deposition, chemical vapor deposition [19], and electrodeposition (wet process) [20]. Single-bath and dual-bath electroplating are the two procedures utilized in the CMMA electroplating process. A single solution comprising all the elements of the metal salts required for deposition in a single bath. Plating current density is varied alternately throughout deposition, with or without variation in mass transfer towards the cathode. In a dual-bath, the substrate is manually or mechanically transported between the two plating solutions, resulting in higher electrolyte consumption as well as the possibility of an oxide layer forming on the cathode surface [19–24].

The current study documents the production of Zn-Fe CMMA coatings on MS employing square pulses from an acidic sulphate bath. The SA was used as a brightener or complexing agent. The corrosion resistance of the coating is compared to that of a single layer coating. The impact of current pulses on coating characteristics and corrosion behavior is examined. The fluctuation of wt. % Fe in the alloy coating is the primary reason for Zn-Fe CMMA coatings with improved corrosion resistance.

2. Experimental method

A flat Mild steel specimen of approximately 3 cm × 2 cm = 6 cm^2 area of the substrate was taken as cathode and a zinc plate of the same area as anode. Sand paper of fine grade was used for surface polishing. Then the polished substrate was rinsing with distilled water and finally cleaned with trichloroethylene at room temperature (30°C). The electrolytic bath was formulated using research grade chemicals and distilled water. Electroplating was dependent on pH and time. The pH increases, the formation of metal hydroxide deposited on the substrate resulting increase in thickness and it is mainly depends on the time for deposition. Other factors like temperature, concentrations of electrolyte, applied potential/current are also influence the morphology, electrochemical behavior, thickness and hardness of the coating. The pH of the plating solution was adjusted to 3.0 by adding required quantity of sulfuric acid. The experiments were carried out under room temperature (30°C) and at a manually stirring with time of 10 minutes for deposition. The coated substrates were cleaned with water after electroplating followed by drying. The electroplating bath compositions and functioning parameters for Zn-Fe monolayer and multilayer coatings were optimized with the normal Hull cell system [1]. Sulphanilic acid (SA) is a colorless, water-soluble compound and also complexing agent with metal ions. Citric acid (CA) serves as a buffer to maintain the pH of the bath and sodium acetate is

Bath components	Amount (g/L)
Zinc sulphate	100.0
Ferrous sulphate	20.0
Sulphanilic acid	1.0
Citric acid	4.0
Sodium acetate	50.0
Functional parameters	
Anode	Pure zinc
pH	3.0
Temperature (°C)	30
Current density (A dm^{-2})	3.0

Table 1.
Solution compositions and functional parameters.

conducting agent. The solution compositions and functional parameters for the electroplating were given in **Table 1** for achieving smooth deposit on the cathode.

An electroplating cell, of 200 ml capacity was used. The monolayer and multilayer coatings were carried out using DC power analyzer for 10 minutes (comparison purpose). Potentiostat (CH-Instruments) with standard three-electrode arrangement was used to carry out electrochemical studies. The calomel (SCE), platinum and MS were used as reference, auxiliary and working electrodes respectively. The corrosion characteristics of the coatings were conducted, in 3.5% Sodium chloride (NaCl) solution by potentiodynamic polarization and electrochemical impedance spectroscopy (EIS) techniques. A potential limit of ± 200 mV from open circuit potential (OCP) was used for polarization study, at a scan rate of 1 mV s^{-1}. The frequency range, from 100 KHz to 10 mHz was used for EIS study with perturbing signal of 10 mV. The thickness of the coating was evaluated by faradays law and cross checked using digital thickness tester. Scanning Electron Microscopy (SEM), Model JSM-6380 LA from JEOL, Japan, in the magnification of 1000X was used to observe the surface morphology, cross-sectional view of multilayer formation and to check the layers formed after corrosion test. The hardness of the coating was measured using a micro hardness meter. The composition of metal contents in the coatings was calculated by using colorimetric method [25]. Zn-Fe CMMA coating with alternatively different compositions was defined as: $(Zn\text{-}Fe)_{1/2/n}$ (where1 and 2 signify the two different switched cathode current densities, and 'n' is the

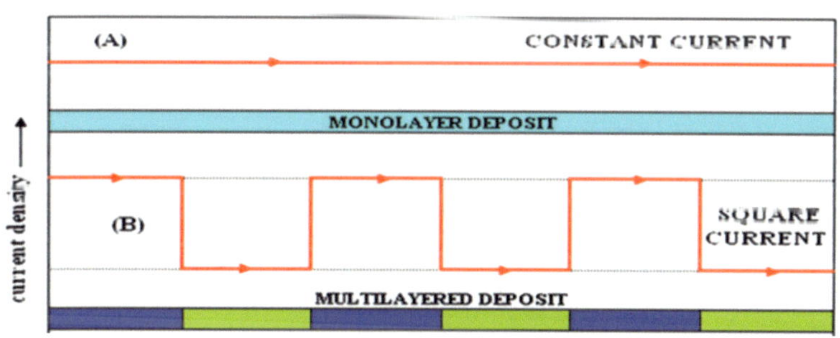

Figure 1.
Schematic representation of direct current pulses (A), and square current pulses (B), respectively used for production of monolayer and multilayer Zn-Fe coating from the optimal bath.

number of layers formed during the total plating period, i.e. 10 min). The Zn-Fe CMMA depositions were produced and characterized with different configurations. All conditions of coatings were kept constant, except current density. Alternatively, programmed cathode current densities were applied by accurate power source set-up. Monolayer and multilayered composition-graded deposits were produced using DC, and square current pulses, as shown schematically in **Figure 1**.

3. Results and discussion

3.1 Effect of current density

The monolayer Zn-Fe coatings have been developed onto MS using direct current from the optimized bath shown in **Table 1**. In the alloy electroplating, the current density (CD) has significant effect on the coating composition, structure, and properties. The effects of CD on metal content, CCE, hardness, thickness, and corrosion resistance of the coatings were reported in **Table 2**. It was found that the iron content in the deposit have been increased with CD (**Table 2**). It is due to the fact that at high CD, more content of Zn deposited on cathode, as observed in Zn-Fe group metal alloys. However, at low CD, the wt. % Fe in the deposit was increase (**Table 2**), may be due to the tendency of bath to follow the normal co-deposition [1]. It could be due to lower hydrogen liberation, which results in lower alkalinization at the cathode film, as explained by the hydroxide suppression process [26]. The cathode current efficiency (CCE) is increased with increase in CD's.

In all CD's of the coatings, the CCE is high (>79). But at higher CD, a small decrease in CCE was observed. This may be due to large evolution of hydrogen (**Table 2**). The hardness of the coatings depends on the applied CD used for deposition. It may be observed that at small CD, less amount iron content in the deposit and showed less hardness. The hardness of the deposit is found to increase with CD, and decreased at higher CD. This may be due to porous metal hydroxide into the crystal structure, caused by much liberation of hydrogen during electroplating. The thickness of the deposit is increased with CD (**Table 2**). It may be due to the adsorbed metal hydroxide, caused by stable increase of pH during plating. The Zn-Fe alloy deposited at the optimal CD (3.0 A dm^{-2}), the deposit shows smooth and uniform coating with lowest corrosion rate compared with other CDs shown in **Table 2**.

3.2 Optimization of CCCD's

The production and characterization of 3D nanostructured crystallites has received the most attention among nanostructure materials, followed by those of

i (A dm^{-2})	pH of bath	Wt. % Fe	CCE (%)	Hardness (VH)	t (μm)	$-E_{corr}$ (V vs. Calomel)	CR × 10^{-2} (mm y^{-1})
1.0	3.0	1.95	79.1	121	9.5	1.013	35.71
2.0		2.37	83.2	138	10.1	1.019	29.09
3.0		3.89	90.8	155	11.3	0.994	20.17
4.0		4.13	87.2	151	12.5	1.015	30.78
5.0		4.95	85.1	147	15.0	1.125	40.29

Table 2.
Effect of current density on deposit characters of monolayer Zn-Fe coating, deposited from optimized bath at 30°C.

SCCD's (A dm^{-2})	-E_{corr} (V vs. Calomel)	i_{corr} (μA cm^{-2})	CR × 10^{-2} (mm y^{-1})
Multilayered Zn-Fe coatings produced at the difference of 3.0 A dm^{-2} between SCCD's			
$(Zn\text{-}Fe)_{2.0/5.0/10}$	1.182	9.99	15.29
CMMA Zn-Fe coatings produced at the difference of 2.0 A dm^{-2} between CCCD's			
$(Zn\text{-}Fe)_{3.0/5.0/10}$	1.177	12.56	18.37

Table 3.
Corrosion rates of multilayer Zn-Fe coatings developed from the optimized bath at different set of SCCD's with 10 layers each.

2D layered nanostructures. The former will be employed in applications requiring high strength, better formability, and soft magnetic characteristics, while the latter will be used in electrical and corrosion applications [27]. Multilayer deposits were formed using the square current pulse approach, which has been demonstrated to improve corrosion resistance in a variety of metals and alloy systems. As a result, multilayer coatings with alternate layers of alloys of varying compositions have been produced. Using the potentiodynamic polarization approach, the corrosion behavior of monolayer and multilayered Zn-Fe alloy coatings with varying numbers of layers is described, and the associated corrosion results are provided in **Table 3**. **Table 3** shows that the coatings produced at a difference of 2.0 and 5.0 A dm^{-2} had the lowest corrosion rate among the various sets tested. These SCCDs were chosen for additional research into the effect of layering, as indicated in the sections below.

3.3 Optimization of number of layers

To achieve a well-defined differentiation between Zn-Fe layers of different composition, Multilayer coating have been produced at 2.0/5.0 A dm^{-2} and 3.0/5.0 A dm^{-2}. The Zn-Fe CMMA coatings having 10, 20, 60, 120, 300 and 600 layers, including 10 layers were developed and their corrosion rates were measured by potentiodynamic polarization method.

Corrosion rates in both sets of SCCDs were shown to decrease as the number of layers increased up to 300 layers, then increase, as shown in **Table 4**. At 2.0/5.0 A dm^{-2}, the corrosion rate reached saturation (beyond which there was no further decline in corrosion rate with layering) at 300 layers with the lowest corrosion rate (0.41×10^{-2} mm y^{-1}), compared to 20.17×10^{-2} mm y^{-1} of monolayer Zn-Fe coatings. Though there is a significant decrease in corrosion rate at 3.0/5.0 A dm^{-2}, as shown in **Table 4**, the data for 2.0/5.0 A dm^{-2} have been considered. However, an attempt to improve corrosion resistance by increasing the number of layers (i.e. 600 layers) in each set of SCCDs resulted in an increase in CR (**Table 4**).

The increased corrosion rate at higher degree of layering (such as 600 layers) is due to a shorter relaxation period for the ion redistribution (Zn^{+2} and Fe^{+2}) in the diffusion layer [2]. Because the overall time for deposition remains constant, the time for deposition of each layer, such as $(Zn\text{-}Fe)_1$, decreases as the number of layers increases. With a high degree of layering, there is not enough time for metal ions to relax and deposit on the cathode, resulting in composition modulation. As a result, compositional modulation at a high level of layering is unlikely to occur. To put it another way, the CMMA deposit is becoming monolayer. Therefore, $(Zn\text{-}Fe)_{2.0/5.0/300}$ has been proposed as the optimal configuration of multilayer coating, with individual layer thickness ~ 50 nm; for peak protection against corrosion.

SCCD's (A dm^{-2})	No. of layers	-E_{corr} (V vs. Calomel)	i_{corr} (μA cm^{-2})	CR (×10^{-2} mm y^{-1})
Optimization of layer thickness at SCCD's of 2.0–5.0 A dm^{-2}				
$(Zn\text{-}Fe)_{2.0/5.0}$	10	1.215	10.01	15.89
$(Zn\text{-}Fe)_{2.0/5.0}$	20	1.220	9.56	14.26
$(Zn\text{-}Fe)_{2.0/5.0}$	60	1.233	6.57	8.93
$(Zn\text{-}Fe)_{2.0/5.0}$	120	1.235	3.16	4.34
$(Zn\text{-}Fe)_{2.0/5.0}$	300	1.261	0.39	0.46
$(Zn\text{-}Fe)_{2.0/5.0}$	600	1.247	8.11	11.98
Optimization of layer thickness at SCCD's of 3.0–5.0 A dm^{-2}				
$(Zn\text{-}Fe)_{3.0/5.0}$	10	1.191	9.54	13.18
$(Zn\text{-}Fe)_{3.0/5.0}$	20	1.195	8.30	12.29
$(Zn\text{-}Fe)_{3.0/5.0}$	60	1.215	5.08	7.87
$(Zn\text{-}Fe)_{3.0/5.0}$	120	1.213	3.43	5.10
$(Zn\text{-}Fe)_{3.0/5.0}$	300	1.181	0.67	0.99
$(Zn\text{-}Fe)_{3.0/5.0}$	600	1.217	8.61	12.78

Table 4.
Decrease of corrosion rate with increase in number of layers of Zn-Fe alloy coatings, produced from the optimal bath.

4. Corrosion study

4.1 Potentiodynamic polarization method

The electrochemical corrosion behaviors of monolayer and multilayer Zn-Fe coatings, having different number of layers are characterized using potentiodynamic polarization method, and results are reported in **Table 4**.

A progressive decrease of corrosion current, i_{corr} and hence, the CR was observed up to 300 layers, after which CR started increasing significantly. The potentiodynamic polarization curves with increase in number of layers as shown in **Figure 2**. From the **Table 4**, it is clear that as increase in the corrosion potential, with decrease of corrosion current density, corrosion rate up to 300 layers and the increased the corrosion rate as explained in 3.3.

The resistance, R, is defined by Ohm's equation, V = IR, as the ratio of voltage V to current I. Its application is confined to the optimal resistor for a DC system that is frequency agnostic. Real electrochemical systems, on the other hand, exhibit far more complex behavior. They are not simply resistive. The capacitive term is added by the electrochemical double layer. Diffusion, for example, is a time and/or frequency dependent electrode process. As a result, impedance is employed instead of resistance in an actual electrochemical system. The AC response of an electrochemical system, denoted as Z(ω), is the AC response of the system under investigation to the introduction of an AC signal (e.g., a sinusoidal wave) [28]. On the basis of electrical double layer capacitance, EIS is a useful approach for gaining vital information about the interface (between substrate and medium), which is responsible for better corrosion resistance [29]. Nyquist plots were used to gain information on the interaction of the coating with the corrosive medium. It was observed that in $(Zn\text{-}Fe)_{2.0/5.0}$ coatings, the radius of the semi-circle increased with number of layers, up to 300 layers as shown in **Figure 2** indicating that corrosion rates had decreased with layering. Because the same bath chemistry and cell

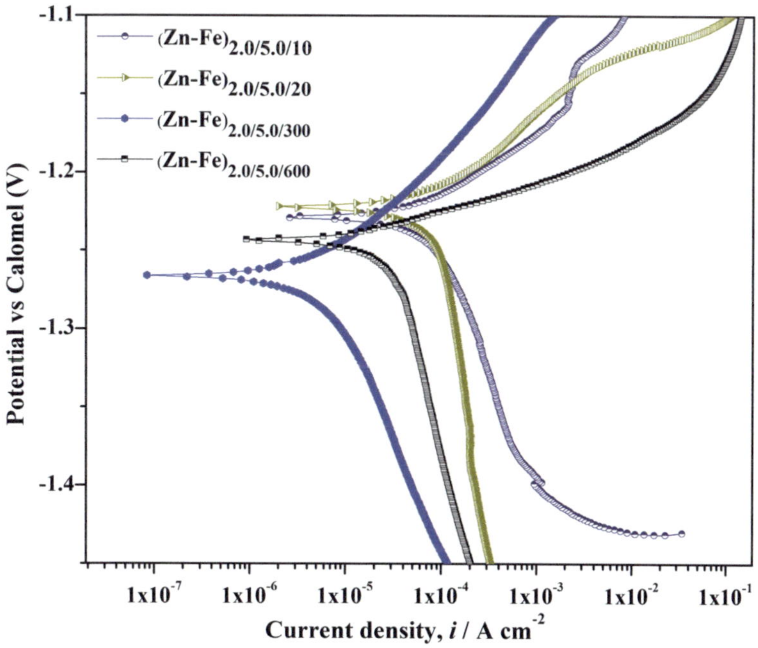

Figure 2.
Potentiodynamic polarization curves of multilayer $(Zn\text{-}Fe)_{2.0/5.0}$ coatings, produced from optimized bath at 30°C with increased number of layers.

configuration were utilized, the solution resistance Rs is essentially comparable in all possible layers. However, as the number of layers increased, polarization resistance, or R_P, increased. The radius of the semi-circle reduced significantly in the case of a deposit with 600 layers (magnified image shown in inset), suggesting poor corrosion resistance as shown in **Figure 3**.

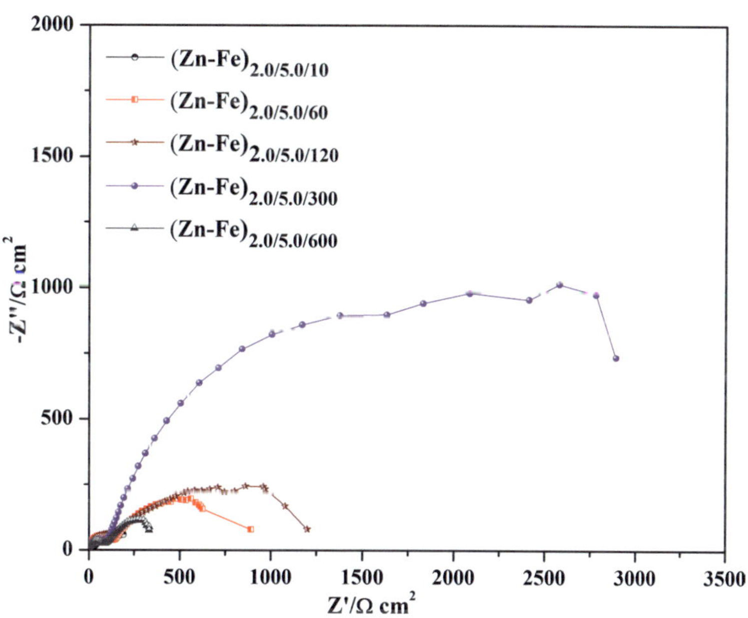

Figure 3.
Electrochemical impedance response of multilayer $(Zn\text{-}Fe)_{2.0/5.0}$ coating produced from optimized bath at 30°C with increased number of layers.

Coating configuration	$-E_{corr}$ (V vs. Calomel)	i_{corr} (μA cm^{-2})	CR$\times10^{-2}$ (mm y^{-1})
$(Zn\text{-}Fe)_{3.0/monolayer}$	0.994	14.71	20.17
Multilayer $(Zn\text{-}Fe)_{2.0/5.0/300/square}$	1.261	0.39	0.46

Table 5.
Comparison of corrosion rates of monolayer and multilayer Zn-Fe alloy coating of same thickness.

4.2 Comparison between monolayer and multilayer Zn-Fe coatings

It may be noticed that i_{corr} value of $(Zn\text{-}Fe)_{3.0/monolayer}$ is much less compared to $(Zn\text{-}Fe)_{2.0/5.0/300/square}$ coatings. The improved corrosion resistance afforded by multilayer coating can be explained in terms of the formation of alternate layers with different phase structures, caused by low and high metal (Fe) content in the deposit. The CR of multilayer $(Zn\text{-}Fe)_{2.0/5.0/300}$ coating configuration is found to be ~43 times less (0.46×10^{-2} mm y^{-1}) than compared to monolayer Zn-Fe coating at optimal current density (3.0 A dm^{-2}) (20.17×10^{-2} mm y^{-1}), obtained from same bath, for same time (10 min). A large change in corrosion rate was observed when coatings was changed from monolayer to multilayer, and is supported by the corrosion data reported in **Table 5**.

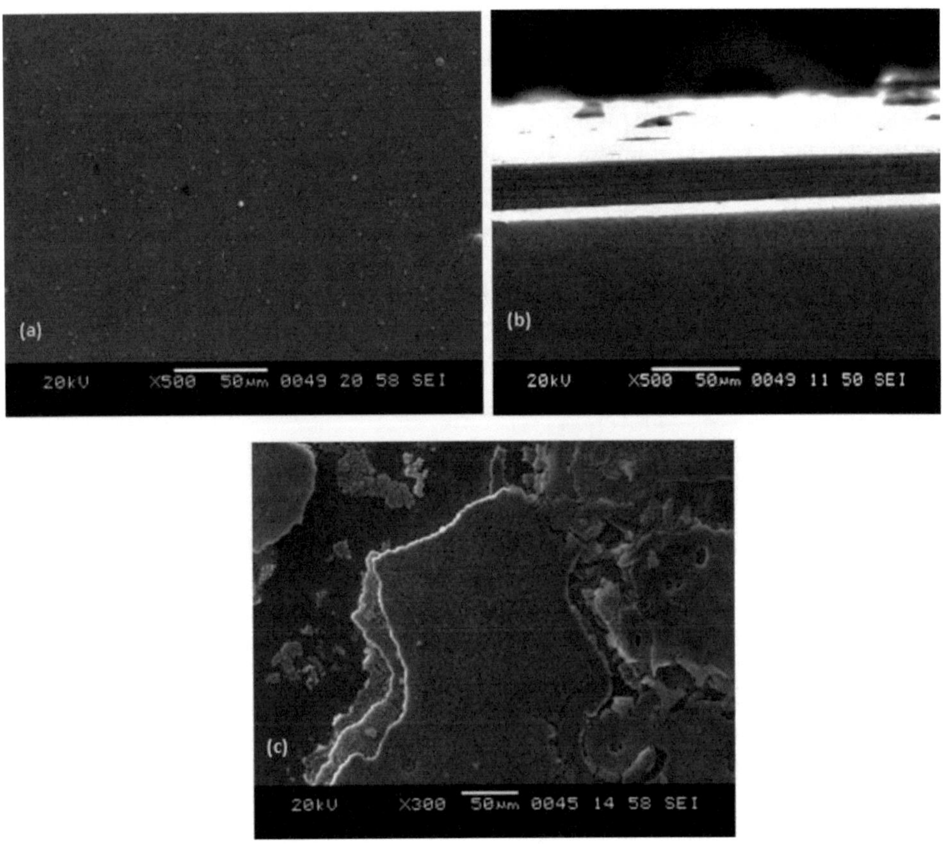

Figure 4.
The surface morphology of $(Zn\text{-}Fe)_{3.0}$ (a), CMMA $(Zn\text{-}Fe)_{2/5/10}$ (b) and $(Zn\text{-}Fe)_{2/5/10}$ after corrosion test.

5. SEM study

SEM images of Zn-Fe alloy having different coating configurations; deposited using direct and square current pulse is shown in **Figure 4a–c**. The surface topography of monolayer Zn-Fe coating at optimized current density (3.0 A dm^{-2}) as shown in **Figure 4a** displayed a smooth and uniform morphology. **Figure 4b** shows the cross-sectional view of multilayer $(Zn\text{-}Fe)_{2.0/5.0/10}$.

The inspection of the microscopic appearance of the surface allows understanding the mechanism of corrosion; with possible reasons for improved corrosion resistance. The high corrosion prevention is due to fact that one layer of alloy having one type of failures (like pores, crevices, due to deposition at one current density) will be covered consecutively by the another layer, having another type of failures (due to deposition at other current density). Thus the coating possesses alternate layers, having different degree of failures, and consequently the corrosion agent's path is longer or blocked. That is why in multilayer coating the corrosive agent needs more time to penetrate through than in case of monolayer coating. As a whole, the protection efficacy of multilayer $(Zn\text{-}Fe)_{2.0/5.0/300}$ coatings may be explained by the barrier effect of Zn-Fe layer, with high wt. % Fe and the sacrificial effect of Zn-Fe layer, with less wt. % Fe (**Table 2**) A small change in wt. % noble metal in the alloys layer is good enough to bring large change in the phase structures of the alloys and thereby their properties [23]. The examination of the acid treated surface under SEM confirmed the formation of layers as shown in **Figure 4c**. To understand the reason for the improved corrosion resistance, the multilayer coatings were examined under SEM after corrosion test.

The 2D and 3D surface images of Zn-Fe monolayer and multilayer alloy coated samples, obtained by AFM measurement are shown in **Figure 5a,b**. The mean roughness (R_a) and root mean square roughness Z_{rms} of the coatings are calculated from AFM images. In monolayer coating at optimal current density (3 A dm^{-2}), the deposit was uniform, and has a huge peak with bigger grains shown in **Figure 5a** and the R_a and Z_{rms} values are 56.8 and 69 nm. Where us in

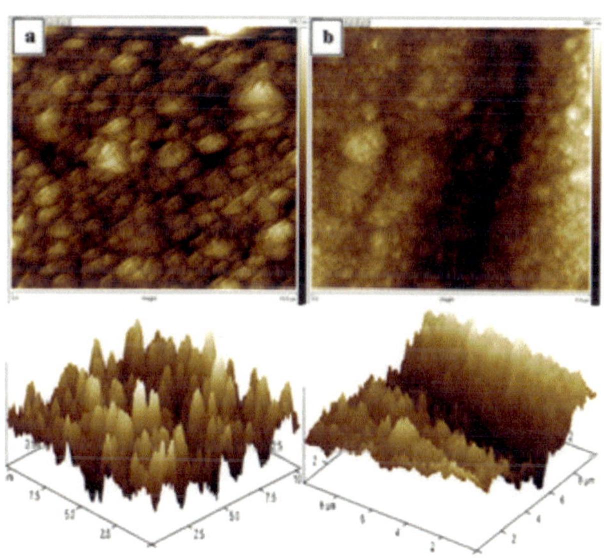

Figure 5.
AFM images of Zn-Fe monolayer (CD, i = 3.0 A dm^{-2}) (a) and multilayer alloy coatings $(Zn\text{-}Fe)_{2/5/10}$.

case of multilayer coating at optimal number of layers (Zn-Fe $_{3/5/300}$), the deposit was uniform in texture, homogeneous, and exhibited small peaks with regular grain size as shown in **Figure 5b**, and R_a and Z_{rms} values are 28.5 and 30 nm and hence multilayer coatings shows better corrosion resistance compared to monolayer coating.

6. Conclusions

The following conclusions were drawn from the above study

- Under optimal condition, multilayered Zn-Fe coatings, developed using square current pulse is about ~ 43 times more corrosion resistance than corresponding monolayer alloy coating.
- The thickness of individual layers play important role in imparting better corrosion resistance. The corrosion rate decreased with number of layers only up to 300 (optimal) layers, and then increased. The increase of corrosion rate at high degree of layering was attributed to less relaxation time for redistribution of metal ions in the diffusion layer.
- The protection efficacy of multilayer coatings is attributed to the difference in phase structure of the alloys in consecutive layers, deposited at various current densities.
- The SEM and AFM study have demonstrated the surface chemistry and topography of alloy coating, with reasons responsible for better corrosion resistance of Zn-Fe alloy coating.
- The corrosion stability of the multilayer coatings is related to the defects and failures occurring in a single layer in the deposition process, can be neutralized or masked, by the sequentially deposited layers. Therefore, the corrosive agent path is extended or blocked.

Acknowledgements

I thankful to the NMAM Institute of Technology, Nitte, Karnataka, India for providing the instrumental facilities.

Conflict of interest

There is no conflict of interest declared by the authors.

Acronyms and abbreviations

CD	current density
AFM	atomic force microscopy
SEM	scanning electron microscopy
SCCD	switched cathode current density
SA	sulfamic acid

CA	citric acid
CMMA	compositional modulated multilayer alloy
CCE	cathode current efficiency
SBT	single bath technique

Author details

Ramesh S. Bhat
Department of Chemistry, NMAM Institute of Technology, Nitte, Karnataka, India

*Address all correspondence to: rameshbhat@nitte.edu.in

References

[1] Brenner A. Electrodeposition of Alloys-Principles and Practice, N.Y. Academic Press. 1963; 1: 2.

[2] Nasser Kanani. Electroplating: Basic Principles, Processes and Practice. Elsevier Ltd, Berlin, Germany, 2006.

[3] Jensen JD, Gabe DR and Wilcox GD, The practical realization of zinc-iron CMA coatings. Surf. Coat. Technol. 1998; 105: 240-250.

[4] Hegde AC, Venkatakrishna K, Eliaz N, Electrodeposition of Zn-Ni, Zn-Fe and Zn-Ni-Fe alloys. Surf. Coat. Technol. 2010; 205: 2031-2041.

[5] Bhat RS, Bhat KU, and Hegde AC, Optimization of deposition conditions for bright Zn-Fe coatings and its characterization. Prot. Met. Phys. Chem. Surf. 2011; 47: 645-653.

[6] Bhat RS, and Hegde AC, Studies on electrodeposited Zn-Fe alloy coating on mild steel and its characterization. J. Electrochem. Sci. Eng. 2019;9: 9-16

[7] Zhang Z, Leng WH, Shao HB, Zhang JQ, Wang JM, and Cao CN, Study on the behavior of Zn–Fe alloy electroplating. J. Electroanal. Chem. 2001; 516: 127-130.

[8] Lan CJ, Liu WY, Ke ST, and Chin TS, Potassium salt based alkaline bath for deposition of Zn–Fe alloys. Surf. Coat. Tech. 2006; 201:3103-3108.

[9] Narasimhamurthy V, and Sheshadri BS, Electrodeposition of Zn-Fe alloy from an acid sulfate bath containing triethanolamine. Plat. Surf. Finish. 1996; 83:75-79.

[10] Panagopoulos CN, Agathocleous PE, Papachristos VD, and Michaelides A, Sliding wear behavior of zinc–iron alloy electrodeposits. Surf. Coat. Technol. 2000; 123:62-71.

[11] Yang CQ, Long ZL, and Zhou YC, A sulphate bath for the preparation of Zn-Fe alloy coatings. Trans. IMF. 2002; 80:161-163.

[12] Diaz SL, Mattos OR, Barcia OE, and Fabri Miranda FJ, Zn-Fe anomalous electrodeposition stationeries and local pH measurements. Electrochim. Acta. 2002; 47:4091-4100.

[13] Thangaraj V, Eliaz N, and Hegde AC, Corrosion behavior of composition modulated multilayer Zn-Co electrodeposits produced using a single-bath technique. J. Appl. Electrochem. 2009;39: 339

[14] Wilcox GD, and Gabe DR, Electrodeposited zinc alloy coatings. Corros. Sci. 1993; 35(5-8):1251-1258.

[15] Bhat RS, and Hegde AC, Production of layer by layer Zn-Fe compositional multilayer alloy coatings using triangular current pulses for better corrosion, protection, Transactions of the IMF, 2015;93(3):157-163

[16] Rahsepar M, and Bahrololoom ME, Corrosion study of Ni/Zn compositionally modulated multilayer coatings using electrochemical impedance spectroscopy, Corros. Sci. 2009; 51: 2537-2543.

[17] Fei J, Wilcox G, Electrodeposition of zinc-nickel compositionally modulated multilayer coatings and their corrosion Behaviors, Surf. Coat. Technol. 2006; 200:3533-3539.

[18] Nabiyouni G, Schwarzacher W, Rolik Z, Bakonyi I. Giant magneto resistance and magnetic properties of electrodeposited Ni-Co-Cu/Cu multilayers, J. Magn. Magn. Mater. 2002; 253:77-85.

[19] Dobrzanski LA, Lukaszkowicz K, Pakula D, and Mikula J, Corrosion

resistance of multilayer and gradient coatings deposited by PVD and CVD techniques. Arch. Mater. Sci. Eng. 2007; 28:12-18.

[20] Leisner P, Nielsen CB, Tang PT, Dorge TC, and Muller P, Methods for electrodepositing composition-modulated alloys, J. Mater. Process. Technol. 1996; 58:39-44.

[21] Jensen JD, Gabe DR, and Wilcox GD, The practical realization of zinc-Iron CMA coatings, Surf. Coat. Technol. 1998; 105:240-250.

[22] Ivanov I, and Kirilova I, Corrosion resistance of compositionally modulated multilayered Zn-Ni alloys deposited from single bath. J. Appl. Electrochem. 2003; 33:239-244.

[23] Prabhu Ganeshan, Swaminatha PK, and Branko NP, Development of compositionally modulated Zn-Ni multilayer deposits as replacement for cadmium. Surf. Coat. Technol. 2007; 201:7896-7904.

[24] Bhat RS, KU, and Hegde AC, layered coating of Zn–Co alloys on mild steel using triangular current pulses for better corrosion protection. T. Indian. I. Metals. 2013; 66:193-199.

[25] Vogel AI, Quantitative Inorganic Analysis (Longmans Green and Co, London), 1951.

[26] Horans J, Effect of plating parameters on electrodeposited NiFe. J. Electrochem. Soc. 1981; 128: 45-49.

[27] Gurrappa I, and Leo B, Electrodeposition of Nano-structured coatings and their characterization-a review. Sci. Technol. Adv. Mater. 2008;9: 043001

[28] Yuan X, Song C, Wang H, and Zhang J, Electrochemical Impedance Spectroscopy in PEM Fuel Cells – Fundamentals and Applications. Springer Publications, London, 2010.

[29] Craig BD, Fundamental Aspects of Corrosion Films in Corrosion Science. Plenum Press, New York. 1991.